Strategic Minerals and International Security

Contributors:

R. Daniel McMichael
Richard E. Donnelly
Paul Krueger
Robert Wilson
Russell Seitz
Sheldon Glashow
Wallace E. Kirkpatrick
E. F. Andrews
James D. Santini
Paul C. Maxwell
Robert L. Terrell
William Schneider, Jr.
John R. Thomas

Editors

Uri Ra'anan and Charles M. Perry

Special Report
On the Agenda: 1
July 1985

A Joint Publication of the
INSTITUTE FOR FOREIGN POLICY ANALYSIS
and the
INTERNATIONAL SECURITY STUDIES PROGRAM
THE FLETCHER SCHOOL OF LAW AND DIPLOMACY

PERGAMON·BRASSEY'S
International Defense Publishers

Washington London New York Oxford Toronto Sydney Frankfurt

Pergamon Press Offices:

U.S.A.	Pergamon-Brassey's International Defense Publishers, 1340 Old Chain Bridge Road, McLean, Virginia, 22101, U.S.A.
	Pergamon Press Inc., Maxwell House, Fairview Park, Elmsford, New York 10523, U.S.A.
U.K.	Pergamon Press Ltd., Headington Hill Hall, Oxford OX3 0BW, England
CANADA	Pergamon Press Canada Ltd., Suite 104, 150 Consumers Road, Willowdale, Ontario M2J 1P9, Canada
AUSTRALIA	Pergamon Press (Aust.) Pty. Ltd., P.O. Box 544, Potts Point, NSW 2011, Australia
FEDERAL REPUBLIC OF GERMANY	Pergamon Press GmbH, Hammerweg 6, D-6242 Kronberg-Taunus, Federal Republic of Germany

Library of Congress Cataloging in Publication Data
Main entry under title:

Strategic minerals and international security.

(On the agenda ; 1 (June 1985)) (Special report / Institute for Foreign Policy Analysis)
Papers presented at a forum held on Sept. 17, 1984, sponsored by the International Securities Study Program of the Fletcher School of Law and Diplomacy, and the Institute of Foreign Policy Analysis.
"Published in cooperation with the Institute for Foreign Policy Analysis."
1. Strategic materials--Government policy--United States--Congresses. 2. Mineral industries--Government policy--United States--Congresses. 3. United States--National security--Congresses. 4. United States--Foreign relations--1981- --Congresses. I. McMichael, R. Daniel. II. Ra'anan, Uri, 1926- . III. Perry, Charles (Charles M.) IV. Fletcher School of Law and Diplomacy. International Securities Study Program. V. Institute for Foreign Policy Analysis. VI. Series: On the agenda ; 1. VII. Series: Special report (Institute for Foreign Policy Analysis)
HC110.S8S78 1985 333.8'5'0973 85-12336
ISBN 0-08-033157-2 (pbk.)

Printed in the United States of America

Contents

Preface

The International Security Studies Program of The Fletcher School of Law and Diplomacy and the Institute for Foreign Policy Analysis, Inc., joined forces on September 17, 1984, as on previous occasions, to address a topic of particularly urgent concern – namely, the security aspects of access to, and utilization of, strategic materials and minerals. This work was made possible, in part, by a generous grant extended by the J. Howard Pew Freedom Trust, as well as by the ongoing magnanimous support received from the Trustees of the Sarah Scaife Foundation. The editors were fortunate to be able to avail themselves of the contributions and the expertise provided by leading members of the U.S. Administration's National Strategic Materials and Minerals Program Advisory Committee.

The gathering responded to the evident need for infrastructure work, in assessing current efforts, governmental and private, to deal with America's vulnerabilities concerning commodities essential for the preservation of national security and industrial competitiveness, as well as currently acceptable standards of living.

The editors found that the contributions made, both in the formal submissions to the Forum and in the ensuing discussions, were of major value and significance, adding substantially to our knowledge of the topic and suggesting further initiatives to address these and related problems.

It is hoped that this publication, the first in the On the Agenda series, will provide a significant contribution to the agenda of America's foreign affairs and defense community, at the initial phase of the second Reagan Administration.

Charles M.Perry
Senior Staff Member
Institute for Foreign Policy Analysis, Inc.

Uri Ra'anan
Professor of International Politics
Director, International Security Studies Program
The Fletcher School of Law and Diplomacy

Executive Summary

It is the hope of the editors and contributors that the materials presented in this Special Report will focus public attention and official action on the importance of planning strategic minerals policy as a key component of our national security and economic well-being. While it was not intended to produce a special set of recommendations acceptable to all, it appears to the editors that certain initiatives are implicit in the analyses presented, and that lively public discussion concerning these and other suggestions may help to further the task of enshrining materials and minerals policy in the appropriate place on the national defense and foreign policy agenda. Among the proposals that deserve closer scrutiny are measures to:

- Broaden public understanding of strategic minerals vulnerabilities and their potential impact on industrial prosperity and national security.
- Reaffirm the value of the National Defense Stockpile, both as a deterrent to the cut-off (or manipulation) of foreign minerals supplies, and as an emergency source in times of war or national crisis.
- Improve the content and quality of the national stockpile through the sale of excess items, the purchase of new materials relevant to current or emerging high technologies, and the location of stocks close to potential industrial consumers.
- Make better use of barter trade as an economical way to acquire commodities for the stockpile that are not produced, or are in short supply, within the United States.
- Consider the creation of a special government-owned corporation or board of experts—similar perhaps to the Federal Reserve Board—that would have sole responsibility for stockpile management.
- Set up an inter-agency task force to identify key sectors in the minerals/materials industry that are becoming obsolescent and may be in need of greater federal assistance, including tax incentives, loans, and purchase guarantees.
- Explore, in particular, the wider application of Title III of the Defense Production Act (DPA) as a means to sustain and stimulate domestic production and processing capabilities that are crucial to the national defense.

- Conduct a full review of the status of public lands to determine their resource content and to define precisely which lands are and are not open to possible exploration and lease.
- Promote further Department of Defense efforts to develop new manufacturing technologies (such as robotics and computer-aided design), as well as synthetic materials (e.g., composites), that may reduce consumption of strategic minerals vulnerable to interruption.
- Continue research on a variety of procedures for processing minerals in outer space, and for the eventual utilization of extra-terrestrial resources (primarily from the moon and near-Earth asteroids).
- Move forward with the development of a continental shelf mining option as an alternative to stalled schemes for deep seabed mining.
- Discourage multinational lending authorities (e.g., the World Bank) from making loans that may lead to overproduction of key commodities, particularly in developing countries, and create an imbalance in world supply and demand, thus rendering unsubsidized production facilities in the United States less competitive.
- Develop a keener appreciation of Soviet "resource diplomacy" and the challenges it poses to import-dependent industrial democracies, especially Japan and the West European countries.
- Encourage greater alliance cooperation in the area of mineral supply security, including joint research and development programs and stockpile coordination (and exchanges), as well as closer collaboration in providing economic and security assistance to key producer countries in the Third World.

Strategic Minerals: The Public Policy Process

*by R. Daniel McMichael**

Mineral resource dependency—particularly in terms of import dependency—as a public policy issue has been an off-again-on-again matter ever since the beginning of World War II. During the 1940s, dependency, as well as other key factors, such as mobilization and preparedness, appeared high on the public policy agenda. Back then, the reasons were clearly perceived by the American public in the context of the goals of the time: first, to win a major war and, then, to "keep the world safe for democracy."

For instance, during the Truman era, the Paley Report assessed materials dependency in considerable detail and became the basic guideline in the formulation of resource policy. And the National Security and Defense Production Acts of 1948 codified into law an official recognition of U.S. dependency on imported minerals and raw materials, with various provisions, particularly stockpiles and standby allocation, designed to safeguard sources of supply during catastrophic emergencies, such as another major war.

During the 1950s, with the Korean War and the Eisenhower era, public policy practices involving resource dependency continued as little more than rote bureaucratic procedures; after all, the geopolitical posture of the United States in those days was largely unthreatened, because the Soviet Union still lacked a significant global reach. In those days, the United States, along with our Western allies, ruled the seas without challenge, a factor absolutely vital to securing sources of minerals supply. The minerals treasure houses of the Middle East and Southern Africa were still comfortably within the scope of Western influences. Thus, public concern about import dependency, understandably, was not all that high.

The 1960s saw us preoccupied with Southeast Asia and our own domestic problems, so that by that time "minerals dependency" was pretty much a

*Mr. McMichael is the Vice Chairman of the National Strategic Materials and Minerals Program Advisory Committee.

non-issue. We still had the rituals of our public laws. We still had a stockpile program of sorts—revised, cut up, reorganized again and again, abused, misused, and not taken very seriously by rank-and-file public policymakers. After all, the United States and its allies still constituted the greatest, undisputed global power combination. So why should we worry about imported minerals supplies?

Of course, there were a few who thought otherwise, such as miners, geologists, and other technical experts, who have been concerned always about the nation's mineral needs. And the technical journals and field reports are full of references to the finite nature of our natural resources. But, save for this rather vertical professional interest, historically there has been relatively little widespread public policy interest—until recently.

The prelude to this emerging interest began during the mid-1960s, in the Defense Department, in a few scattered think tanks, and among a handful of Chief Executive Officers, when concern began to grow regarding a subject this country had never had to consider much before—dependency on foreign supplies of minerals and raw materials. This would have sufficed, I suppose, if the United States had remained *the* undisputed global power of the 1940s and 1950s; but, as we all know, times were changing, moods were changing, and the balance of power was changing. Thus, in the mid-1960s, a number of studies were undertaken and a few conferences and seminars were held—all beginning to look at a whole range of "what if" scenarios.

Then came the 1973 oil embargo, and suddenly one did not need to hold six degrees to comprehend what "minerals dependency" meant. But there was a problem. The public understood oil dependency, because it made *direct* use of petroleum. But, to quote former Congressman James D. Santini, "You don't buy cans of cobalt off the grocery shelf."

The policymaking community and public opinion both still had difficulty identifying non-fuel minerals import dependency as a serious national problem, which could threaten not only our national security but our industrial and technological base as well. Mindful of this void in public thinking, a group of 16 nationally recognized experts came together in Pittsburgh, Pennsylvania, in June 1980, to study a whole spectrum of domestic, foreign policy, and defense-related issues, which in one way or another affected minerals dependency, especially with regard to non-fuel minerals.

Partly as a result of that study, Congressman Santini convened hearings in September 1980. The issue was picked up by the American Geological Institute, which submitted a statement of concern to all of the presidential candidates. Ronald Reagan responded, with the support of 18 members of the House and Senate, by creating a Strategic Minerals Task Force which functioned during the transition period.

During its first term, the Reagan Administration again responded to these concerns by incorporating strategic minerals dependency as part of the agenda of the President's Cabinet Council on Natural Resources and the Environment. Since then, the Congress and, in particular, the National Security Council and the Departments of the Interior, Commerce, State, and Defense, along with the Federal Emergency Management Agency (FEMA) and the Committee on Materials (COMAT), in the Executive Office of the President, have been active on a number of policy fronts regarding the formation of a minerals policy in keeping with the new requirements of the 1980s and beyond.

Given this background, a brief summary will indicate why minerals security is such an important issue and deserves to be the subject of serious consideration.

The U.S. industrial infrastructure, defense capabilities, and economic viability rest on adequate supplies of more than two dozen non-fuel minerals. Of these, the United States currently is more than 90 percent import dependent on 13 and more than 50 percent import dependent on an additional 13. Table 1 gives key examples to illustrate how just seven of these minerals—in which we are *over* 90 percent import dependent—are absolutely indispensable to our industrial infrastructure.

Combine these elements of foreign supply with those concerning our energy needs, and the cumulative effect of all our mineral import dependencies has the potential of becoming possibly the greatest direct threat to our national interests since the end of World War II—a view held by a number of recognized experts, and one which I share. We reach this conclusion because this form of resource dependency renders the nation highly vulnerable to power projection by competing states and to other circumstances that either have occurred or may occur outside the traditional limits of our sovereignty. This vulnerability derives from the threat of supply cutoff. Whether sudden or gradual, any significant interruption of vitally needed minerals from foreign sources would lead to equally significant disruptions across the spectrum of American life. These would adversely affect national security, industrial productive capacity, and the economy (e.g., income, jobs, inflation, living standards). Obviously, the degree of such supply interruptions would determine the severity of domestic disruptions. But we must anticipate disruptions, the consequences of which could, in turn, adversely affect the U.S. position as a world power.

Table 1.
Key Minerals in Which the United States is Over 90% Import Dependent (1978)*

Indispensable Minerals or Ores Critical to High Quality in the Production of:	Bauxite 93%	Chromium 92%	Cobalt 97%	Columbium 100%	Manganese 98%	Platinum-gr. 91%	Tantalum 97%
Basic steel					x		
Stainless, tool, alloy steels		x	x	x	x		x
Basic aluminum	x						
Aluminum alloys	x			x	x		
Nickel-based and cobalt-based superalloys	x	x	x	x	x		x
Items Made Using Above Basic Products:							
Ordnance (tanks, fighters, bombers, missiles)	x	x	x	x	x	x	x
Power generating (turbines, controls, transmission)	x	x	x		x		
Electric motors, equipment (locomotives to mixers)	x		x		x		
All electronics (appliances, computers, phones, TV, navigation, controls, telecommunications)	x	x	x	x	x	x	x
Nuclear applications	x	x	x	x			x
Jet engines, gas turbines (hot parts)		x	x	x		x	x
Batteries, fuel cells					x	x	
Aerospace (airframes, hulls, rocket engines)	x	x	x	x		x	x
Projectiles, gun barrels, machine parts, crankshafts, axles, gears, machine tools		x	x	x	x		x
Mining, drilling (valve stems & systems, drill bits)		x	x		x		
High-tech medical (cryogenics, heart-lung, scanners)	x	x	x			x	x
Petroleum processing, drilling		x	x	x	x	x	x
Chemical processing		x	x	x	x	x	x
Glass products			x			x	
Pharmaceutical production						x	
Food processing, enrichment		x	x		x	x	
Synfuel production		x	x	x	x	x	x

Note: Since effective military technology requires highest possible performance, substitutes are not desirable. However, substitutes for purely commercial and civilian items are possible on a mixed basis – recognizing lower quality and performance would occur.

Source: *U.S. Bureau of Mines

As mentioned, we already have had one experience with a cutoff which, indeed, created severe disruptions to our economy and life-styles—the 1973 oil embargo. The subsequent cartel success of OPEC left us with some pretty unpleasant economic dislocations with which we are all too familiar—elements of which still plague us today.

The pain of that experience has led to some careful examinations by a growing number of specialists as to what might happen to us were another embargo—or supply cutoff of some kind—to hit us unexpectedly, this time involving critical non-fuel minerals. One recent study, for instance, looks at a worst-case scenario involving a sudden loss in foreign supplies of chromium, absolutely essential to the making of stainless, tool, and alloy steels in all of the derivative products shown in Table 1.

The key scenario in this study—which is built on sophisticated economic The key scenario in this study—which is built on sophisticated economic plier; cuts off all chromite exports to the United States and other Western nations; and also disrupts half of Zimbabwe's production of chromite. Moreover, the USSR and Albania, in this scenario, refuse to make sales to the West, so that the combined result is a 65 percent cutoff of chromite supplies to the non-communist industrialized world. Under such circumstances, the study concludes, the United States would suffer the loss of at least 1.05 million jobs for a period of five years. This is a conservative estimate.[1]

The evidence provided by Table 1 is conclusive: Without the minerals it needs, the United States cannot maintain its industrial infrastructure and thereby its economic strength; and without the latter the nation cannot hope to retain its geopolitical strength and independence—which (to close the circle) it must have in order to ensure its access to foreign supplies of minerals. Thus, the United States must deal with a circle of vulnerabilities that link together virtually every basic component of its domestic and international life. If this circle is broken outright or badly frayed along its course, disruptions in the national life will result that, if allowed to persist, could do serious damage to U.S. vital interests.

These aspects are fairly well perceived by the nation's leaders, insofar as oil dependency is concerned. Given OPEC's objective to control oil prices and production, and given the continuing Middle East crisis, the linkage between our foreign and domestic policies and assured access to imported oil is now well understood. There is, on this subject at least, a general awareness that the nation cannot pursue an effective energy policy without integrating key levels and sectors of government activity—domestic, defense, foreign—and

---

[1] See James T. Bennett and Walter E. Williams, *Critical Issues. Strategic Minerals: The Economic Impact of Supply Disruptions.* (Washington, D.C.: The Heritage Foundation, 1981), pp.47-54.

that any major break in the circle of vulnerabilities, at home or abroad, can cause severe disruptions.

Not so well understood is the fact that the same phenomenon holds for non-fuel minerals. As we review the vulnerabilities in question, it will become apparent why it is essential to seek remedial action.

Central to all discussions concerning strategic minerals is the matter of "free access" to foreign supplies. It is important that we define what is meant by this: namely, a climate that permits enterprises in the United States to bid and participate freely, openly, and equitably in international settings for the minerals essential to the nation's needs – whether these enterprises are linked to defense procurement, or the mining and related processing industries, or end users such as the steel, aluminum, aerospace, or electronics industries. Thus, the issues to be examined here revolve around what, where, when, and how the nation is to preserve free access to foreign supplies of the minerals that are needed to preserve our industrial and technological base – and thereby preserve our life-styles and, most crucial of all, our jobs.

A number of recognized experts conclude that key U.S. foreign vulnerabilities concerning access to strategic minerals at present lie with the following:

- *The Middle East.* The vulnerabilities are obvious.

- *The whole of Southern Africa.* Without free access to this region, absolutely critical supplies of manganese, cobalt, chromium, and platinum-group metals could be denied to the United States with devastating effects on its industrial infrastructure. Southern Africa can scarcely be regarded as a stable region.

- *The Central America/Gulf of Mexico/Caribbean region.* Beyond Jamaica as a major supplier of bauxite, it is the sea lanes that are critical to the national interest. Some 75 to 80 percent of our imported critical minerals are transported through these waters to *fixed* facilities along Gulf and East Coast ports. Thus, they can become vulnerable to interdiction, for instance, by a strong Soviet naval presence in Cuba, the Caribbean, and South Atlantic and/or other destabilizing influences, such as the widening of the conflict in Central America.

- *Soviet influence.* The USSR must be regarded as a hostile power which is developing the capability, through policies and actions, to impede access to international natural resources by the United States and its allies. Soviet statements concerning the USSR's global strategy are consistent on this point, as are many of its actions. With regard to Soviet influence, it is noteworthy that the United States suffers the highest degrees of resource dependency in the Middle East and Southern

Africa, and serious shipping vulnerabilities in the Caribbean, Gulf of Mexico, South Atlantic, and Indian Ocean—the very areas over which the Soviet Union is seeking to establish hegemony, via Moscow's own and proxy forces in Cuba, El Salvador, Nicaragua, Angola, Ethiopia, Yemen, and Afghanistan. The Kremlin is also deploying a four-ocean navy, more and more capable of interdictions. Moreover, the Soviet Union's minerals trading pattern is changing in a way that could endanger Free World access, i.e., by importing certain strategic minerals for itself and Soviet bloc allies from countries whose output traditionally has gone mainly to the United States and its allies.

Must the United States, and its West European and Pacific allies, be forever dependent—and thus vulnerable—on imported sources of minerals, both fuel and non-fuel? Is there no way we can escape the dependency trap? Probably not. At least not entirely. However, that does not mean that this country—with the cooperative support of both the public and private sectors—should not be doing all within its power to lessen its vulnerabilities.

One obvious way, of course, is to diminish our dependency on imports. In the case of oil, we are trying to conserve and to develop various alternative sources of energy that would leave us less reliant on the Middle East as a principal foreign source of supply. In the case of strategic non-fuel minerals, a number of steps are being taken under the Reagan Administration (some having begun under the Carter Administration) and in the Senate and the House of Representatives.

Essentially, Congress and the Executive Branch are trying to hammer out a set of strategic non-fuel minerals policies, designed at least to alleviate our very high degree of dependency on foreign supplies. In addition to work going forward inside Interior, Commerce, State, Defense, FEMA, COMAT, the National Security Council, and the Cabinet Council, three new entities recently have been created either through Executive or Congressional actions: (1) the National Strategic Materials and Minerals Program Advisory Committee, established by (then) Secretary of the Interior William P. Clark; (2) the Public Lands Advisory Council, under the Bureau of Land Management (both of these draw heavily on the private sector); and (3) the National Materials Council, established under the Glickman/Fuqua Bill, signed by President Reagan. Formation of this three-person commission now is in process, with appointments expected soon. So far, these collective efforts within the government have been focused largely on five basic questions:

- Stockpiling of critical natural resources, including materials acquisition and distribution, as well as stockpile management itself, which is a particularly sensitive and difficult problem.

- The fate of the Defense Production Act, especially with regard to the future funding of Title III and, if funds are provided, for what purposes they are to be utilized.
- The broad and complex issues involving the use of public lands, especially as they relate to the exploration, leasing, and development of minerals, including energy resources.
- The development of materials and processing technology, showing promise in ceramics and powdered metals.
- The development of seabed mining. (This does not include *deep* seabed mining, since the Reagan Administration has withdrawn the United States from the Law of the Sea Treaty, largely because currently there is no adequate legal basis in the treaty to encourage U.S. companies to invest in the high costs of sophisticated exploration, mining ships, and processing facilities ashore.) The President, early in 1983, issued a proclamation establishing a new Exclusive Economic Zone for the commercial development of natural resources in an area 200 miles seaward from all U.S. territory. This is a good first step toward establishing a solid, workable base of legal protection, from which U.S. enterprises can begin serious efforts to mobilize capital investment and technology in an effort to reclaim at least a few of the strategic minerals from the seabed.

Not all of these efforts have been harmonious, and various shades of opinion are apparent in the policymaking community. However, the point is that, during the past four or five years, the issues concerning resource dependency—and the vulnerabilities associated with our reliance on imported natural resources to sustain our way of life—have made it to the national policymaking agendas of both the Executive and Legislative branches of government and have been carried more and more into the policy circles of industry and its technical support systems.

This is only a beginning. Much more needs to be done. In the matter of stockpiles—how to build, manage, and use them—confusion still remains. Various elements in the Executive and Legislative branches have offered a wide range of options on how to institute necessary reforms. At this time, a workable consensus has not been reached. One can only hope that sometime reasonably soon a relevant and viable stockpile management program can be brought into being with the support of both the Congress and the Executive.

In the matter of funding Title III of the Defense Production Act, it seems to me that in its deliberations the Congress would benefit from a comprehensive assessment and reorganization of the stockpile program, as the first order of business in grappling with the problem of our resource import dependency. Resulting priorities based on critical need and the degree of import dependency seem to many of us to be the basic criteria upon which Title III funding judgments should be made. In any case, Title III ought not to be lost as an important public policy instrument to help meet critical national minerals needs.

With regard to the use of public lands, critical need and degree of import dependency, again, would seem to be the basic criteria upon which to make these lands available for fuel and non-fuel minerals exploration, leasing, and development. Continued pressure to restrict such access to the public lands needs to be carefully weighed against these criteria, as Congress and the Executive deliberate the pros and cons of this very sensitive issue.

In regard to the newly established Exclusive Economic Zone, too little notice has been taken by policymakers—both public and private—of this significant new development in terms of what it could mean in broadening our access to natural resources. I would hope much more serious attention will be paid to the promises and prospects which the EEZ offers our nation.

Finally, there is the continual need to integrate our defense, foreign, and domestic policies with realistic assessments of our minerals needs—particularly as far as the high degree of import dependency in vital natural resources is concerned. There has been some progress in this area through ongoing efforts in various branches of the Executive, and Congress has demonstrated a growing interest. What was begun in 1980 to clarify some of the issues of dependency and vulnerability needs to be continued. Progress so far has been good, and the response to the contents of that 1980 study constitutes encouraging testimony that the process of integrating our minerals needs with our national and international policies is well in place—and is expanding to the benefit of our overriding national interests.

U.S. Policy and Its Implementation

*by Richard E. Donnelly**

Current initiatives in the Department of Defense with respect to strategic minerals can be traced back to 1980-1981, when several reports were issued calling for a wholesale reassessment of material supply policies. Notable in this regard was a report by the House Armed Services Committee (HASC) entitled *The Ailing Defense Industrial Base: Unready for Crisis*. As shown in Table 2, the report included a number of materials-related statements, along with the finding that the United States lacked an effective capability to handle non-fuel minerals issues. About the same time, the Defense Department was conducting its own study with the Defense Science Board (DSB). This report came out in 1981 with conclusions and recommendations (see Table 3) similar to those of the HASC study, one of the more important of which was the need to support actions by the Federal Emergency Management Agency (FEMA) to utilize Title III of the Defense Production Act (DPA) as a means to strengthen domestic mining and mineral processing capabilities.

Yet, even with the "big ticket" budget that the Defense Department now has, there remain significant resource dependency and vulnerability problems. Foreign dependency applies not only to ores, but to the capability to process ores. If the supply of chromium were cut off tomorrow, not only would the United States face a serious problem in regard to imports, but it would confront as well serious bottlenecks in trying to process effectively the chromium that is stored in the national stockpile. A key question is, would the chromium stocks actually be ready for use by American industry?

As Table 4 illustrates, the Department of Defense has its own "endangered species" list for materials. The snail darter is not on it, but many other equally exotic items are. As an example, consider UDMH, which stands for unsymmetrical dimetholhydrazine. UDMH is a product that had only one producer in the United States, and that producer shut down because of the impact of regulations designed by the Occupational Safety and Health Administration

---

*Mr. Donnelly is the Director for Industrial Resources, Office of the Under Secretary of Defense for Research and Engineering.

Table 2

Major Findings of the House Armed Services Committee

Defense Industrial Base is Deteriorating
- Skilled Manpower Shortages
- Increasing Dependence on Foreign Sources for Critical Raw Materials
- Capital Investment Constrained by Inflation, Unfavorable Tax Policies, MGT Priorities

DOD Lacks Adequate On-going Program or Plan to Address Industrial Base Issue

U.S. Lacks Effective National Non-Fuel Minerals Policy

Contracting Procedures are Excessively Restrictive—Inhibit Stability, Discourage Capital Formation and Reduce Savings

No Focal Point for Leadership—Congress and Executive Branch

Source: U.S. House of Representatives, Committee on Armed Services, *The Ailing Defense Industrial Base: Unready for Crisis*, Ninety-sixth Congress, Second Session, December 31, 1980 (Washington, D.C.: U.S. Government Printing Office, 1980).

Table 3

Key Recommendations of the Defense Science Board Study

Modify Current Legislation, Regulations and Practices to Permit Greater Use of Multi-Year Contracts

Encourage Industry Investment—Expedite Paying Cycle, Increase Use of Milestone Billings, Support Executive Branch and Congressional Actions to Stimulate Capital Investment

Increase Emphasis on Manufacturing Technology and Upgrading of Government-Owned Machine Tool Base

Restructure Current Industrial Preparedness Planning Program

Increase Priority for WRM Stocks & Spares

Place Emphasis on Proper Application of Defense Priority System

Support Actions (by FEMA) to Utilize Title III of the DPA

Source: U.S. Department of Defense, Office of the Under Secretary for Research and Engineering, *Report of the Defense Science Board 1980 Summer Study Panel on Industrial Responsiveness*, January 1981 (Washington, D.C.: U.S. Governnment Printing Office, 1981).

Table 4

Endangered Species:
Key Materials and Products that Could Be in Short Supply

Precision Miniature Bearings	Rubber Footwear	Industrial Fasteners
Jewel Bearings	Ultra-Fine Cobalt Powder	Low Noise Bearings (Subs)
Specialty Metals	Precision Components and Timing Devices	Aerial and X-ray Film
Aerial Refueling Hose	Aircraft Clocks	Rayon
Chaff	High Purity Silicon	Acrylic Sheet
Fine Wire Mesh Filters (Aircraft)	Cathode Ray Tubes (B&W)	UDMH
500 Gal Fabric Drums	Thermal Batteries	Vacuum Tubes
Hydraulic Drive Motors (Sonar Systems)	Large High Quality Forgings	TWT Tubes
		Tubeaxial Fan (MK87 Gun System)

Note: This list does not reflect any official DOD supply priorities. It is only meant to illustrate the type and variety of defense-related commodities that, in certain crisis scenarios, could be in short supply, as a result of foreign source vulnerabilities or domestic production shortfalls and bottlenecks.

Source: Office of Industrial Resources, Office of the Under Secretary of Defense for Research and Engineering.

(OSHA) and the Environmental Protection Agency (EPA). Almost overnight, therefore, the DOD went from spending 50 cents a pound of the taxpayers' money for UDMH to $9.50. We had to drain unsymmetrical dimetholhydrazine out of certain operational missiles for use in other defense-related tasks. The point is that any number of material supply problems can suddenly emerge from a strictly national security standpoint, as well as from a fiscal or readiness standpoint.

After the DSB and HASC reports came out and some of the initial work was done by the transition team for the first Reagan Administration, the Pentagon came up with a plan to improve industrial responsiveness for the national defense, and it should be understood that the question of materials supply was in the forefront of that particular program. While it is not possible here to review all the steps that the DOD is taking, I do want to stress that access to materials is viewed by top defense officials as a key component to improving our national security and overall well being.

In 1982, the President issued a national security directive that called, among other things, for greater efforts in the stockpile area and in using the Defense Production Act (DPA) to ease our resource dependency situation. Also in 1982, release of the President's National Materials and Minerals Program Plan led to the creation of a Cabinet Council on resource issues, together with a Committee on Materials and an Emergency Mobilization Planning Board (EMPB). The EMPB is currently looking at the adequacy of the stockpile, as well as America's ability actually to process stockpiled materials, if there is a need to do so. Within DOD, we have begun assessments of several materials, such as titanium, beryllium and chromium, but we are worried also about a number of the more high-technology materials. Very clearly, the foreign source dependency situation is not good for the items described in Mr. McMichael's paper, but there are some less widely known commodities—UDMH and PAM (polyacrynitrile), for example—the supply of which can be just as serious a concern.

The point also needs to be made that the industrial base improvements initiated by the Defense Department sometimes work at cross purposes. On the one hand, concerted efforts are being made to reduce acquisition costs, raise efficiency, and improve economies. But, on the other hand, more money should be spent to retain or expand our capability for surge production in an emergency; and buying a variety of stocks to put on the shelf, or developing incentives to set up a new industry sector in the United States, is not without cost. Furthermore, it may *not* reduce acquisition costs of weapons and equipment. So there is a careful—if elusive—balance that must be reached in the purchase of war reserve stocks and raw materials, as opposed to finished military hardware for immediate use.

A key initiative of the Defense Department is to do more in the materials research and advanced technology arenas, and to improve the capability of our manufacturing base to stimulate new demand for materials. Referring briefly to Table 5, some of the major "mission areas"—such as strategic defense and the space programs—should be compared with the technology needs listed in the second column, and the programs that DOD has developed so far to address special materials needs. We are not yet ready to have a tupperware aircraft engine, but a large number of very interesting commodities are being developed, and quite a few contain basic ores.

The point ought to be made as well that in addition to the single source problem for raw materials, we also are experiencing several production constraints in the industrial sectors that *use* strategic materials. For example,

Table 5
DOD Materials and Structures: Science and Technology Program

Mission Areas	Technology Needs	Thrusts
STRATEGIC OFFENSE		
Reentry Vehicles Propulsion Systems	All Weather Capability Maneuvering Capability Efficient Rocket Nozzles	Carbon/Carbon Composites Metal-Matrix Composites
SPACE		
Satellite Structures Propulsion Systems Mirrors-Optical Structures Antennas	Survivability Outgassing Thermal/Electrical Conductivity Dimensional Stability High Stiffness	Metal-Matrix Ceramic Matrix Composites Carbon/Carbon Composites
LAND WARFARE		
Tanks Vehicles Mobility	Improved Armor Gun Barrel Erosion Ground Vehicle Survivability	Metals, Ceramics, Organics Metal Matrix Composites
AIR WARFARE		
Aircraft Tactical Missiles	Durability of Composites High Strength "Forgiving" Metals Long Life High Temperature Gas Turbine Components All Weather Capable Seeker Domes	Organic Matrix Composites Metal Matrix Composites Ceramic Matrix Composites
NAVAL WARFARE		
Mines and Torpedoes Ships Survivability Submarines	High Strength "Forgiving" Metals Composites Joining Techniques	Metals Metal Matrix Composites Welding
RESEARCH		
	Understanding Structural Reponse Energy Interactions Synthetic New Materials	Micro/Macro Mechanics Fracture Mechanics

today there is only one viable producer in the United States of armor plate for tanks. The other producers are bankrupt, or about ready to go under. Thus, in addition to having a strong stockpile, we must strengthen the capability of American industries to use materials that would be released from the stockpile; and this is no easy task when 65 percent of the new machine tools purchased for the one cannon manufacturing facility now operating in the United States was produced in foreign countries. There is a need, therefore, to monitor more carefully the resources that are placed in U.S. arsenals, in order to ensure a reasonable balance between reliance on foreign and U.S. supplies and capabilities.

It is worth mentioning, moreover, that the danger of overdependence on imported goods is even worse among our major allies, who have shown little interest in stockpiling. Japan, fortunately, is looking into the idea of expanded stocks for a few materials. The European trading partners, however, are only beginning to think about stockpiling, and to my knowledge, little, if any, inventory is currently in place.

Research and development on composites, carbon carbon, rapid solidification techniques, and laser-hardening represents the bulk of DOD's work in the field of materials science. Progress in all of these areas will affect strategic and critical material availability, and obviously may alter our materials shopping list. For example, in the composites area, in 1972, when the F-15 tactical aircraft was being fielded, 2 percent of its weight was composite material. Today, the AV8B, which is another tactical aircraft, contains 26 percent composite materials by weight. The Advanced Tactical Aircraft, which is not flying

Table 6
Composite Structures in Navy and Air Force Tactical Aircraft

Aircraft	Contractor	Pounds Composite	Percentage Structural Weight	Percentage Structural Weight Saved	First Production Date	Number Produced To Date	Future Production
F-14	Grumman	180	0.8	0.4	10/78	470	375
F-15	McDonnell	218	1.6	2.0	12/74	705 819*	747 790*
F-16	General Dynamics	180	1.8	0.7	4/78	1050*	2600*
F/A-18	McDonnell	1219	10.0	3.9	10/78	97 113*	1280 2459*
AV-8B/ GR-MK5	McDonnell	1430	26.8	8.3	10/83	4	328 546*

*Includes foreign military sales

yet, will be 33 percent composite materials by weight. So there are some important changes taking place in aircraft construction which will help reduce overall requirements for certain critical materials (e.g., cobalt).

Greater efforts also must be made to render U.S. industry more competitive in terms of technology. Toward that end, there is a growing manufacturing technology program within the Defense Department, largely as a result of initiatives taken by the Reagan Administration transition team in 1980. We are now spending about $200 million a year on developing new and improved manufacturing techniques, processes and equipment; and while manufacturing technology primarily is developed to support the needs of the Department of Defense, it has a spin-off into the American commercial market place. Consider near net shape forging—that is, forging a part closer to the ultimate shape in which it will be used. By forging it closer, one can save on the amount of strategic and critical materials consumed, save on machine time, and get production rates up.

It is true, of course, that many of the manufacturing technology projects sponsored by DOD—such as computer-aided manufacturing, computer-aided design, and robotics—have been implemented overseas more rapidly than they have in the United States. However, there is a company with high productivity that produces arc welding equipment. Its return on equity is 19 percent; its management philosophies, the welfare of its employees, and its concern for customers are interrelated; and the average salary is $44,000 per year. Promotion from within is stressed; they have a fetish for cleanliness; and there are no reserved parking places. Wouldn't you like to work there? Well, if you want to see this company, you can sell your ticket to Japan, because this company is in Cleveland, Ohio. The point is that while the United States can learn from its competitors, it also can catch up with them, and do things just as well.

DOD stockpile concerns are addressed in another paper. A couple of key defense policy decisions, however, have had a major impact on current stockpile policy. The decision, for example, to plan for the full spectrum of potential conflict, instead of planning only for a highly intense short duration emergency, has translated into an increased need for raw materials and industrial base planning. DOD also has pushed for a review of the quality of materials stocked and ways to improve it. In particular, there is a need to sell off materials that definitely are excess to requirements, and to use some of the dollars so raised, not only to procure new materials, but to take the materials already stocked and put them into a high state of readiness. So, too, commodities should be stocked which are further along in the manufacturing chain and need, therefore, less energy input to get into a more readily usable form.

Table 7

Purposes/Use of Title III of the Defense Production Act (DPA)

1. Reduce Vulnerability to Foreign Sources
2. Reduce the Need to Stockpile
3. Provide Flexibility over Stockpiling
4. Counteract Soviet Influence to Destabilize Sources/Markets
5. Generate and Retain Investment and Employment in the U.S.
6. Expand Domestic Mobilization Base

Note also should be made of the Defense Production Act (DPA), especially Title III, which was used successfully in the 1950s to provide incentives for domestic production in the materials sector. Recently, DOD has begun to rely once again on Title III programs, but the results have been modest. There were a number of persons, of course, who viewed DPA Title III as the path to an Industrial Renaissance, going so far as to propose that $6.75 billion of the Defense Department's budget be spent on Title III initiatives over four or five years. We at DOD, however, balked at such grandiose plans, arguing that the advantages of spending some six and a half billion dollars to buy raw materials were not compelling, when serious problems remain in "filling the bins" with weapons and equipment needed by the armed forces. Fortunately, after all the noise was over, there was a two and a half year extension on the DPA and an authorization for appropriations of $25 million in 1985 and $75 million in 1986. What we have in mind at DOD is to do our best to move forward with a few Title III projects to confirm the utility of the concept to the general public and Congress, and then, we hope, in later years, to make greater use of Title III authority.

Finally, with the help of the Defense Economic Impact Modeling System (DEIMS), DOD now has the capability to take the five-year budget plan and break it down in such detail that, if we are going to spend a million dollars on a tactical missile, we now know where that million dollars goes into the economy as far as material demand is concerned—so many dollars for aluminum castings, semi-conductors, or whatever commodity you want to consider. Thus, we can now help a company that, for instance, wants to get into the aluminum business, by at least giving our best guess as to what the demand for that product will be over each of the next five fiscal years. We can also make assumptions as far as what the demands from the commercial sector will be at the same time. Consequently, the idea at the Department of

Defense, through techniques like DEIMS and manufacturing technology and several other programs, is to create a demand for special materials within the economy, and to stimulate more capital investment. This may lead to a draw-down on current supplies of strategic and critical materials, but it also should prompt the development of additional domestic resources.

The Stockpile and Emergency Planning

*by Paul Krueger**

Before considering the stockpile, the organizational history of the Federal Emergency Management Agency (FEMA), and an outline of its role in national emergency planning, should be addressed. FEMA was formed in 1979 in an effort to pull together the various emergency functions which were scattered throughout the government, including emergencies relating both to national security and to the more common natural disasters. FEMA is composed, therefore, of five different bureaucratic ingredients: the Defense Civil Preparedness Agency, which was in the Department of Defense; the Federal Preparedness Agency, which was in the General Services Administration; the Federal Disaster Assistance Administration and the Federal Insurance Administration, both of which were lodged in the Department of Housing and Urban Development; and the U.S. Fire Administration, formerly based in the Department of Commerce. My particular office—resource preparedness—dates back to the National Security Act of 1947 which, in addition to establishing the National Security Council, the Joint Chiefs of Staff, and the Central Intelligence Agency, also established the National Security Resources Board. In essence, the Office of Resource Preparedness within FEMA is the successor organization to the National Security Resources Board, and it is concerned primarily with ensuring that sufficient supplies of resources are available during an emergency of any kind.

A good deal of our work is concerned with raw material resources and, in particular, with the national defense stockpile. The history of the stockpile itself goes back to the 1880s when the United States Navy converted from wooden ships to ships of iron and steel. It was recognized then that many of the materials which went into these ships were not readily available within the United States, and some of these commodities—manganese, especially—remain a problem today. In any case, although the Senate held hearings in 1886 on foreign resource dependency, nothing really was done other than to bring the issue to the surface. This was the case also shortly after World War I

*Mr. Krueger is the Assistant Associate Director for Resources Preparedness with the Federal Emergency Management Agency.

and in 1930, when Congress held additional hearings on resource supply problems. In 1938, however, Congress finally appropriated money for the Navy to build up strategic resource stockpiles; and shortly thereafter, in 1939, the Strategic and Critical Material Stockpiling Act was passed, which provided the federal government, for the first time, with statutory authority to acquire vital materials and to put them into a national stockpile.

Six materials were acquired prior to the start of World War II, and for several items, these stocks were the only supplies available to the United States during the war. In 1946, based upon the wartime experience of coping with shortages in raw materials supplies, Congress amended the Stockpiling Act of 1939, and the national defense stockpiling program was begun in earnest. In the early 1950s, given the Korean War, a tremendous amount of material was acquired in the space of a few short years, valued at about two and a half billion dollars in 1950 prices. At the same time, 8 billion dollars worth of loans and purchase guarantees were provided under Title III of the Defense Production Act, in an effort to bring on additional domestic resources. These incentives, by the way, allowed us to create a titanium industry where none had existed before, to increase copper capacity by some 25 percent, to start a domestic nickel industry, and to double the production of aluminum, all in the space of 4 to 5 years. In addition, about 38 billion dollars worth of tax incentives were offered, through rapid write-offs on new investments. Since then, however, little has happened.

Most of the materials that are now in the stockpile were acquired during the 1950s, and we have let stockpile management slide. Some materials in the stockpile were used during the Korean War, and others were utilized during the Vietnam War, but little has been done in recent years to replenish or expand the stockpile. Fortunately, however, there was a major reassessment of stockpile needs in 1976, and we are in the process of trying to implement the recommendations of that study. The basic philosophy of how to use the stockpile has remained the same since 1976, but we did update stockpile needs in 1980, and there is another study now underway to update stock requirements once again.

All of this activity has taken place under the Strategic and Critical Materials Stockpiling Act, which specifies that stockpiles should be sufficient to meet the military, basic industrial, and essential civilian needs of the country for not less than three years during a national emergency. The stockpile management responsibilities outlined under that act are lodged primarily with FEMA and the General Services Administration (GSA). FEMA is responsible for designating which materials are to be kept in the stockpile, how much of each material should be stored, what quality material should be acquired, and what items ought to be placed on the annual shopping list of what to buy and sell. For its part, the GSA is responsible for the actual acquisition of all stockpile materials, for disposing of materials that have been determined to be in excess of current needs, for the ultimate storage of materials currently in stock, and for quality control.

Now what about the current composition of the national defense stockpile? At last count, there were some 61 basic classes of materials. The present inventory is valued at about 12 billion dollars, and current needs for the total stockpile are estimated to be in the range of 20 billion dollars. Thus, the stockpile remains about 8 billion dollars short. This really does not tell the full story, since some of the existing inventory is larger than necessary. There is, in fact, about 4 or 5 billion dollars worth of current inventory that could be sold off. It must be done, however, in a way that does not disrupt commercial markets.

What we are considering, then, and what we ought to be doing, is selling off that excess inventory, taking the money so earned, and converting it into needed raw material stocks. FEMA and GSA are the principal agencies assigned responsibility to take these steps, but they cannot act alone. There are four other agencies which have a strong impact on stockpile programs, most notably the Department of Defense, which is responsible for helping to establish overall defense needs for a number of hypothetical conflict scenarios, and especially for identifying particular material needs in cases where DOD is a major user and where its uses are abnormal compared to other industrial uses. To cite an example, the economy as a whole uses a considerable amount of tungsten, primarily in light bulbs and in metal-cutting, metal-forming machines. The Department of Defense, in addition to these applications, also uses tungsten in armor-piercing projectiles. Obviously, during any sort of military scenario, DOD's use of tungsten would rise significantly, and the amount that would be needed cannot be estimated by some fancy econometric model based upon peacetime consumption patterns. This is a case where FEMA must go directly to DOD officials and work with them to come up with a figure for how much of the item in question would have to be stocked for defense needs.

A second major contributor to stockpile planning is the Department of Commerce, which establishes patterns of industrial usage to help FEMA determine industry requirements. In addition to determining how industry uses a particular commodity, Commerce officials help to establish quality specifications for the materials to be purchased for storage in the stockpile. These estimates are related as well to supply projections developed by a third agency, the Department of Interior, which helps FEMA and GSA to determine what supplies actually are available within the United States and overseas. Finally, the fourth major player in stockpile policy is the Department of State, which is responsible for assessing the political reliability and economic stability of foreign suppliers, particularly in the event of an emergency that might close off traditional sources of supply. Other groups, such as the National Security Council, the Office of Management and Budget, and the Department of Treasury, also have an interest in the stockpile, but the four agencies just described are the principal members of the stockpile community.

Now to the question of stockpile planning itself. In recent years, we have identified—in good measure through the help of public interest groups and research organizations outside government—that there is, in fact, a strategic materials problem, and that the stockpile is one way to mitigate that problem. Within the last couple of years, moreover, money has been appropriated or provided to acquire additional stockpile materials. In 1984, for the first time, Congress went beyond what the Administration requested, and it now appears that there will be 185 million dollars in fiscal year 1985 for stockpile purchases.

In this context, a few points should be made with respect to the kinds of commodities that need to be acquired. In general, the stockpile as it now stands—most of which was acquired during the 1950s—is adequate for the bulk materials, which have a wide variety of fairly ordinary uses. This would include such items as lead, zinc, tin, and aluminum. Where the stockpile tends to be inadequate is in the higher technology materials. Our first order of business, then, has been to buy greater amounts of materials such as cobalt and titanium, which go not only into the more sophisticated weapons systems, but also into industrial systems that operate at high temperatures for relatively long periods of time. In addition, a new item recently has been added to the stockpile list—namely, germanium. In the past it was not necessary to stockpile germanium, as there were sufficient quantities readily available in the United States to cover essential requirements. However, due to changes in industrial practice, and to the development of new applications, particularly in the area of infra-red optics, there is a need to have on hand larger inventories of germanium for emergency use.

Of course, in addition to the stockpile, there are several alternative methods of dealing with materials supply problems. There is, for example, the Defense Production Act (DPA), which contains two provisions—Title I and Title III—that are directly relevant to material/mineral supply. Title I of the DPA allows for the setting of supply priorities and allocations prior to an emergency. It provides, in effect, a mechanism whereby critical Department of Defense and other national security projects are given priority access to the market place, when such access is needed to carry out the national security mission. There are two ways to do this. One is through the DOD contracting process, which provides for what are called "rated orders," that pass all the way down through the contracting chain. The second method is to use the Defense Materials System, whereby the Department of Commerce—working with the Department of Energy and the Department of Defense—goes out to the major manufacturers of steel, nickel-base alloys, copper-base alloys, and aluminum, and pre-allocates a certain portion of each firm's production to meet national security needs. This approach has been relied on routinely for more than 30 years, so it stands as a proven strategy for ensuring that national security needs are met on a daily basis.

Title III of the DPA also has been used with remarkable success, primarily in the early 1950s, as another way to improve supply security. Efforts now are being made—largely in DOD—to reactivate Title III, and use it once again as a way to strengthen domestic mining and processing capabilities. As a result, the defense budget will include 25 million dollars for Title III action for fiscal year 1985 and 75 million dollars for FY 1986.

Beyond these measures, the stockpile itself can be drawn down for emergency supplies when they are needed for national security purposes. On 29 occasions since the 1940s there have been "Presidential releases" from the stockpile. The most recent was in 1979, when a special grade of asbestos was needed for a critical weapons system. Since that particular grade of asbestos was no longer available in the commercial market, it was released from the stockpile and transferred to the Department of Defense, which then supplied the asbestos to the appropriate defense contractors to keep production of the weapon on schedule.

The biggest problem with the stockpile—and, for that matter, with the general area of industrial mobilization—is the conflict between national security goals and economic goals. This has been the case throughout the history of the stockpile, and it remains the case today. One of the more troubling issues in particular is that, unless there is a clear and pending threat, the country has not been able to take action to address material supply problems in a serious way. We took concerted action during the early 1940s and again in the 1950s, but we have had little experience in industrial mobilization since then. Yet, much of the basic industrial infrastructure was created in the 1950s, and it is now becoming obsolete. Many of our steel and copper facilities—what are called the heavy industries—were built 30 years ago, and we still are living off that investment. At the same time, we are witnessing a fundamental change in the structure of the U.S. economy. We are moving away from heavy industry into the higher technology industries. We are also moving away from manufacturing into sales and services, and that is good for the economy and provides new jobs. However, this trend is not necessarily good for national security; and if we expect industry to make a national security contribution during an emergency, we may have to spend, subsidize, and otherwise provide incentives at the federal level in order to insure that the capability we need, or perceive we need, to mobilize our industrial reserves is there when we need it. It must be borne in mind that there may be considerable conflict between national security and economic priorities, and that national security may not necessarily be economical. To the extent that the nation understands and accepts this potential discrepancy, however, we will have taken a major step toward insuring that adequate mineral and material supplies will be on hand to cover demand in any foreseeable crisis.

Natural Resources: Dependency and Vulnerability

*by Robert Wilson**

Concerning the twin issues of dependency and vulnerability, one fundamental point needs to be established – namely, that with respect to natural resource supplies, dependency *is* vulnerability. Import dependence, in turn, is affected by two relatively simple factors: one is geography, and the other, the relative economic advantage of one producer country over another. Geographical factors can be discussed and dismissed very simply – either we have minerals on the shores or in the territories of the United States, or we do not have them. If we do not have a deposit, but need one, we will have to rely on imports. That is not to ignore the fact that there are quite a few land restrictions on Federal lands, which make it difficult to know what we do and do not have on shore. But the primary focus of this paper is on the economic advantages that lead to import dependency.

Economically caused dependency can be dealt with by attacking the basic problems (e.g., high domestic mining costs, restrictive land-use regulations) and/or by providing for alternative supply arrangements and contingencies. By that I mean such options as the maintenance of a mineral/materials stockpile for national emergency. We have to look as well, however, at some of the economic reasons for the decline of the defense industrial base – if there is, in fact, a decline.

Economic dependency, as most interested observers are aware, can be explained by the Ricardo theory of comparative advantage, which is really "the cheaper-the more efficient" theory: If your country can produce goods more cheaply and efficiently than my country, then you should produce them. The frightening aspect, however, is that overall trends in the raw material sector suggest that almost all basic raw materials and commodities can be produced more cheaply abroad than they can in the United States. The implications for national security are staggering, particularly in light of Dr.

---

*Mr. Wilson is the Director of the Office of Strategic Resources in the U.S. Department of Commerce.

Thomas' paper, which suggests that the Soviet Union probably is looking at the political aspects of resource dependency more closely than we are.

The plight of our basic domestic industries is caused by their loss of advantage over foreign competitors. Moreover, the reason for that loss is generally higher U.S. production costs. Let me refer interested readers to a book published in 1980 by the World Affairs Council of Pittsburgh under the title *Resource War in 3-D*, which has an excellent article by Murray Weidenbaum, former Chairman of the Council of Economic Advisers. Mr. Weidenbaum describes the higher production costs of the United States in detail, and attributes them primarily to higher regulatory costs, as well as to higher energy costs and, in some cases, to higher labor costs.

Of particular importance to the topic of import dependence and vulnerability, is the question of how to make domestic industries, particularly the mining industries, more competitive. Mr. Andrews' paper points out the basic dilemma with reference to specific strategic minerals, including some of the "exotic" materials, such as chromium, cobalt, and manganese. But it may not be too long before we face similar difficulties with regard to such basic commodities as copper, lead, or zinc. We do have deposits of these materials in the United States, but the increased operating costs of American companies, compared to foreign competitors, may drive U.S. mining activities further offshore. A mining company representative, during the recent copper debate, made a point which really hit home; he said it seems fundamentally unfair that the U.S. government in the past has imposed huge burdens on domestic industry through higher taxes, regulatory costs, and wages. Not that some of these costs are not proper, but after they are imposed, we just shove U.S. companies out into the international arena and say, "Compete fairly with your foreign rivals."

The situation becomes even more ridiculous when one considers the fact that the U.S. government, though not at the initiative of the Reagan Administration, actually has subsidized mining ventures in other countries—the rationale being to diversify overall sources of supply. Thus, on the one hand, American industries have to compete while bearing the burden of extra costs imposed by government policy, and, on the other hand, companies overseas do not have to carry this same burden and, in fact, may be indirectly subsidized by the American government through U.S. participation in the International Monetary Fund and World Bank. Truly, there is a problem here, and it must be addressed. Nevertheless, before Americans rush to embrace protectionism, it should be recalled that industrial competitiveness, or comparative advantage, merely means that someone can produce at lower cost.

To regain advantage, either the American side will have to lower its costs, or the other side will have to raise its costs. Now, I am not sure to what extent

Chile can be made to raise its costs. Maybe Chilean copper producers should be subject to the same type of environmental restrictions that industries face in the United States. To the extent that the world is one eco-system, that argument may be made. Yet, it is hardly likely to happen, unless we are able to have the regulations of the Environmental Protection Agency (EPA) apply to Chile. In any case, U.S. costs may be completely justifiable in terms of U.S. socio-economic goals, while foreign producer costs may be unjustifiable by any reasonable standards. I will not discuss the particulars for copper, but it looks as if these issues are pushing the United States toward another battle of priorities on environmental questions.

Of course, we all want clean air, we all want clean water, we all want wilderness lands to save for posterity. By the same token, however, we want to have industries that can compete and land that is in the productive mainstream of the economy. The battle of priorities over the next ten years, particularly in regard to the mining industry, is going to be fascinating. How do you get, for example, the American Mining Congress and the Sierra Club to sit down at one table, making them both realize that the issue of mineral supplies is important and that the national security is affected? To the extent that all sides will come together and talk, maybe there is a chance for more balanced land policies. For, as Mr. Santini points out in his paper, mining operations have upset no more than .02 percent of the U.S. land area. The mining companies are not out there digging up America and leaving it in such poor condition that future generations will not be able to enjoy the land.

In the meantime, however, what do we do? President Reagan has developed a plan for mineral and materials supply, which was sent to Congress in April 1982. The cornerstone of the Reagan plan is free enterprise, and as a preview of how it might work, I refer you to a paper by Dr. Henry Nau that originally was distributed by the Council on Economics and National Security.[1] The paper concentrates on the energy crisis and how the United States reduced its oil vulnerability, but the analogies to non-fuel minerals problems are clear. There are some differences, of course, yet the cornerstone of the energy policy was to free up the marketplace.

The second priority of the energy program was to build up the strategic petroleum reserve, which has its parallel in what President Reagan has done to strengthen the non-fuel stockpile. Indeed, the fact that we made any purchases at all, while faced with the tightest budget deficit conditions in the history of the country, shows a commitment to resolving the supply vulnerability issue. This commitment, moreover, has even weathered pressure from

---

[1]This paper has since been published under the title, "U.S. Energy Security Policy: Finally on Track," in *Geopolitics of Energy*, September 1984.

the Office of Management and Budget (OMB), which generally has opposed further stockpile expenditures.

Under the President's plan, we have also devoted more money to research on alternative materials, and we are trying to improve the coordination of existing development programs. It is important as well to remember that President Reagan is the first President in 30 years to address seriously the minerals problem. From the 1980 campaign, to Mr. McMichael's transition team task force, to the April 1982 policy statement, to the new advisory committee reporting to the Secretary of Interior, and to the National Critical Materials Council to be set up at the White House, this Administration has taken action. To be sure, there are a lot of problems yet to address, and there is going to be a national debate over the question of priorities. Nevertheless, we are making strides in trying to get "from here to there."

Meanwhile, the problems of comparative advantage remain. But before we rush out to embrace protectionism, it should be realized that there is on the books legislation (such as Section 232 of the Trade Act of 1964) that has been used to curtail imports when they threaten national security. But absent that threat to national security, there has been a great deal of emphasis placed on raising productivity and efficiency within the U.S. economy. Both the Secretary of Commerce and the President have criticized American industry when they thought it was necessary, urging U.S. companies to become more productive and more efficient. We are looking through the anti-trust statutes now, trying to determine whether or not they should be changed to allow mining or processing companies to cooperate more closely on basic generic types of R&D. And we are looking as well at whether or not participation in the international lending agencies is justified, and at the possibility of putting more "strings" on the loans that are made. A little bit of "hard-nosed" negotiating on these matters would definitely benefit the United States.

So, too, the Defense Production Act (DPA) has been reauthorized. We now have some 25 million dollars in the defense budget that can be used, *inter alia*, for loan guarantees and financial assistance to sustain the existing surge capacity of vital industries. Reagan Administration officials also think that we now have more balanced land use policies than existed in the past. Probably more important than anything else, however, has been the regulatory reform movement and its task force chaired by Vice President Bush some years ago. It was not the Administration's goal to tear down all the pre-existing regulations; rather, the goal was to make sure that any new policies would go through a rigorous cost-benefit analysis, so that when acid rain or superfund waste cleanup issues are brought to national attention, the price tag and job impact of proposed solutions are carefully considered.

In conclusion: To the extent that our import dependency is due to geographic factors, we have little control over the causative features, although we do need more realistic land-use policies. On the other hand, to the extent that dependencies are caused by economic factors, we have collectively the means to reduce disadvantageous costs. Yet, in reducing those costs, there may have to be some trade-offs between commonly agreed-upon social and economic goals, trade-offs about which none of us really can complain. For example, free and unfettered trade is our goal, but when national security is threatened by imports, government must step in to provide some relief and guidance.

Over time, it is hoped that efforts to restore economic advantage to the United States will nip some problems in the bud before they grow more serious. For we have always treated the mineral and materials issues as a subset of the economy; as an old cliché says, “a rising tide lifts all boats,” and to a certain extent this is indeed true. If the economy is growing stronger, it would seem axiomatic that each segment of it should also be strengthened. To be sure, there are some problems in the mining industry, but the Reagan Administration is giving them special attention. In the absence of immediate national security problems, however, the marketplace and the private sector must take the lead, as it has in the oil and gas sectors.

The Needs of Science and the Stockpile

by Russell Seitz and Sheldon Glashow* **

Russell Seitz

Within the scientific community there is growing interest in a new technology and material—gallium arsenide—a remarkable semi-conductor that is expected to constitute the basis of important sectors of high technology in the near future. It has received some hundreds of millions of dollars of research support over the years, and is certain to play a role in the next generation of electronics, both military and civilian. Yet, the element gallium is relatively new to the stockpile and there seems to be little interest in significant purchases. One of the difficulties, moreover, in obtaining commitment to stockpile evolution that is sensitive to changes in technology is the perception that idle capital (i.e., stocks) somehow is not to be treated with as much respect as capital devoted to more useful ends. This attitude contrasts sharply with certain remarkably sophisticated aspects of the Swiss national stockpiling program. In addition to having availed themselves prudently of large quantities of industrial commodities, the Swiss have also made the most of the opportunity of deploying stockpiled material in ways that are useful to science. Instead of sitting idly, their railway steel, for example, is stored as part of the passive gamma ray shielding for the European accelerator at the European Center for Nuclear Research (CERN).

A similar approach might be used to insure proper gallium storage, for one of the most intriguing developments in physics has been the use of gallium as a detector for neutrinos (it is perhaps a hundred thousand times more efficient than any other material) to ascertain the answers to fundamental problems concerning the way in which the sun works, or where matter may (or may not) be in the universe. Because of the necessities of the stockpiling program, an opportunity may exist to effect the implementation of a remarkable experi-

---

*Mr. Seitz is Director of Technology Assessment for R. J. Edwards, Inc.

**Dr. Glashow, Higgins Professor of Physics at Harvard University, received the Nobel Prize for Physics in 1979.

ment and to secure cooperation from scientists in a joint venture which could, at one and the same time, assist physics at a much lower level of expense and help build the stockpile. One should remember that during the copper shortage of World War II, hundreds of tons of silver were lent by the Treasury to build transformers for the Manhattan Project. Perhaps more importantly, every dime was paid back.

Professor Sheldon Glashow

As a theoretical physicist, I would like to comment on the gallium experiment. It was realized many years ago that the sun probably is operating by nuclear fusion; hence, there should be a copious source of neutrinos, with about a hundred billion of them coming to earth on each square centimeter per second, day and night. It is important, therefore, to detect these neutrinos; and this has been done by Ray Davis and his collaborators at the Brookhaven National Laboratory, in an ingenious experiment for a specialized component of the neutrinos coming from the sun, the most high-energy neutrinos—the ten parts per million that possess energies of more than ten million electron volts. This experiment required many tons of a fluid, wherein the chlorine would be chemically changed by the neutrinos.

In the course of this experiment, it was discovered that the neutrino flux from the sun, particularly in the case of the high-energy neutrinos, varies by a factor of three from what could be expected according to our theoretical understanding of stars. There is no known explanation at the moment for this variation. Consequently, one would like to design an experiment that could detect all of the neutrinos from the sun, and this might be done by involving gallium which is incomparably more sensitive to the particular neutrinos in question. Such an experiment was proposed as a joint venture, involving German, Israeli, and American scientists, and prototype studies have been made with small amounts of gallium. However, the price of the material was rather high at the time, and the total cost would have amounted to $30,000,000, since 50 tons of gallium are required. The Department of Energy rejected the experiment because of costs. Consequently, a new approach is under consideration—comprising the Soviet Union and the Federal Republic of Germany, with about twelve tons of gallium; the Japanese have also expressed interest in this experiment. If strategic stockpiling of gallium were to prove to be of national importance, we physicists could contribute to the costs while doing our work which, incidentally, is nondestructive of the gallium itself.

Comments from the Floor

The problem is that there are no current plans to purchase gallium, because it has not been identified as a strategic material within the meaning of the act under which the national stockpile operates. Consequently, at best, acquisition could prove to be such a protracted process that it could not possibly succeed in time to be of help to the physicists. Material in the stockpile may be used solely for national security purposes, and scientific or budgetary considerations would not be relevant under the current legislation.

Some ten or twelve tons of gallium constitute the total of the national inventory and are almost entirely within the private sector. Perhaps, therefore, the problem could be left for private initiative to resolve. However, the use of derivatives in military electronics may increase to such an extent that, by 1995, approximately 100 tons of gallium per year may be required, a pace of growth roughly parallel to that of silicon in semi-conductors in the 1960s and early 1970s. There is, moreover, no mechanism now in place to warn private industry adequately of impending surges in the defense sector's demand for gallium or, for that matter, any other high technology materials currently not stockpiled. For these reasons, a re-examination very well may be required of our approach to the stockpile and gallium acquisition.

Response to Comments by Russell Seitz

Evidently there remains a problem of communication between the government's research and development departments and those responsible for industrial preparedness, such as the administrators of the stockpile program. The expenditure of hundreds of millions of dollars over a decade to develop the electronic materials upon which the next generation of semiconductor technology will depend should have led automatically to an investigative effort by FEMA and other concerned federal agencies. Nevertheless, there are no current plans for acquiring gallium.

Who knows, moreover, how many other materials—recognized by those involved in state of the art programs sponsored by the Defense Advanced Research Projects Agency (DARPA) as being critical to evolving high technologies of military importance, but unknown to a smokestack-industry-oriented stockpile program—may fall by the wayside, unrecognized until it is too late to accelerate easily their production to meet breakout demand? Better liaison between FEMA and the most sophisticated sectors of the scientific community is needed to avoid future demand crises in regard to high technology materials.

Lunar and Asteroid Materials: Potential Applications and Recovery Methods

*by Wallace E. Kirkpatrick**

In any assessment of mineral supply issues, it is important to consider the potential utilization of materials contained in lunar soils and near-Earth asteroids. To be sure, the current state of knowledge concerning extra-terrestrial resources is rather limited. However, sufficient data does exist to indicate significant potential deposits that could be used both in space-based operations as well as on Earth.[1]

The Moon

Extensive sampling of lunar materials was conducted at the six Apollo landing sites. Examination of these samples and associated geophysical data has provided both detailed knowledge of the surface materials to be found at these sites, and a basis from which to extrapolate with regard to other lunar regions. The rocks of the lunar surface contain up to 18 percent by weight aluminum, up to 20 percent silicon, and up to 46 percent oxygen, and can be mined readily for all three commodities. Iron is present in metallic form, mostly as iron-nickel alloy, in concentrations of a few tenths of one percent in all lunar soils. This iron is easily extracted in the laboratory with an ordinary hand magnet.

Lunar soils also contain glassy fragments (agglutinates) that incorporate metallic iron, which would have to be separated before it could be used as such. So, too, lavas from the maria—the large volcanic craters on the moon's surface—are rich in chemically combined iron (up to 17 percent by weight), and titanium is present in some lunar lavas in concentrations as high as 7.8 percent. Dense titanium minerals from lavas or related materials may have

*Mr. Kirkpatrick is the President of DESE Research & Engineering, Inc.

[1]The potential defense applications, recovery methods, and enabling technologies for the use of near-Earth resources were addressed in a workshop conducted by the California Space Institute in August 1983. The results were published in a document entitled, *Defense Applications of Near Earth Resources*, Cal Space Ref No. CS183-3, revised edition dated October 31, 1983.

accumulated into even richer ones. Silicon is abundant in all lunar rocks and soils, and comprises about 21 percent by weight of most soils. Oxygen, not available in significant amounts as the free element on the moon, is the major constituent of most lunar materials. It can be released from silicate by several methods, including direct electrolysis.

It has been noted by several researchers that the first step in the production of the above materials in many cases involves the heating of lunar soil. This drives off hydrogen (50 parts per million), nitrogen (100 parts per million), carbon (100 parts per million in combination with oxygen or hydrogen), and argon (1 part per million), which are all useful byproducts, or even primary products, if their value warrants heating large amounts of soil to produce them. Also, lunar silicates can be converted into glass and ceramic products.

The lunar samples in our possession come from nine small sites within a very restricted region of the moon's nearside. Orbiting X-ray and gamma ray experiments sensed only 10 percent of the farside surface. The X-ray experiments indicated concentrations, or relative concentrations, of silicon, magnesium, and aluminum averaged over regions of 400 or more square kilometers, while gamma ray operations identified concentrations of thorium, iron, magnesium, potassium, and titanium averaged over regions of 2500 or more square kilometers. Such spatial resolution is too broad for the detection of specific ore bodies. Thus, lunar samples still provide the only real measurements—and the only data of high spatial resolution—for most chemical elements.

The Near-Earth Asteroids

There are millions of small, solid bodies orbiting in the solar system, ranging in size from numerous microscopic fragments all the way up to moon-like spheres, hundreds of kilometers across. Most of these bodies follow orbits that keep them between the paths of Mars and Jupiter. While these asteroids may someday become valuable resources, the Earth-approaching asteroids, specifically those which are more accessible in the near-term, are more germane to this discussion.[2]

Approximately 75 near-Earth approaching asteroids (NEAs) with determined orbits are presently known, and about 20 of these are poorly to moderately understood insofar as their physical characteristics are concerned. Current data allow estimates of size, major mineral components, and approximate thermal inertia at surface, and it has been verified that quite a few asteroids

---

[2]See John S. Lewis, Department of Planetary Sciences, University of Arizona, "Exploitation of Space Resources" (unpublished manuscript).

PERGAMON PRESS INC.

INTERNATIONAL PUBLISHERS OF SCIENTIFIC BOOKS & JOURNALS
NEW YORK • OXFORD • TORONTO • SYDNEY

ARMY MAGAZINE
Book Review Editor
2425 Wilson Blvd.
Arlington VA 22201

We are pleased to send to you enclosed a complimentary copy of the following book for review:

Strategic Minerals and Internati
Edited by Uri Ra'anan & Charles M. Derry
ISBN 0 08 033157 2:softcover $9.95
Published October 1985

1 Copy 033157 2 to be sent by 01/10/85
50/R

Published/Distributed by
☐ **Pergamon Press Inc.** ☐ **Pergamon Press Canada, Ltd.**

Please send two copies of your review to:

Manager/Publishing Services
Pergamon Press Inc.
Maxwell House, Fairview Park
Elmsford, New York 10523

Printed in U.S.A.

are meteorite-like in mineral assemblage. It appears that some have surface coverings of fine particulate material up to a depth of one centimeter. Others are unfragmented rock. In either case, most appear to consist of unaltered rocky crystals at the surface. Close to (or on) their surfaces, a number of near-Earth approachers appear to have readily accessible metallic deposits, commonly referred to as "free metal."

The composition of asteroids is known only from their reflection spectra, but information on the composition of meteorites appears to be roughly applicable to that for the NEAs studied so far. Similar to their meteoritic spectral analogues, the C (carbonaceous) asteroids can be expected to contain up to 10 percent water, 6 percent carbon, several percent sulfur, and useful amounts of nitrogen. The S-type or "strong" asteroids, more common near the inner edge of the main asteroid belt, may contain 10 to 30 percent free metal (iron-nickel alloys with high concentrations of precious metals). Metallic (M) asteroids may be nearly pure metal. What follows is a list of strategic materials that one could expect to find on near-Earth asteroids.

Rhodium (C)
Ruthenium (C)
Palladium (C)
Gold (C)
Platinum (C)
Diamond (U)
Vanadium (E)
Chromium
Cobalt (C)
Manganese (E)
Copper (C)
Iron (C)
Nickel (C)
Tungsten (C)

Legend: C = Carbonyl product from metal phase
E = E chondrite or E achondrite source
U = unreilite source

In addition to these observations, the following points are worth noting:[3]

1. The Earth is gravitationally downhill from the moon and the asteroids. It is highly likely, therefore, that relatively small machines deployed on these bodies could eject toward Earth-space steady or packaged flows of mass much larger than those which could be sent from Earth by systems of similar size. Consequently, the transport infrastructure needed to supply a given flow of materials from the moon or asteroids could be relatively small and modestly powered. Compared to Earth experience, the amount of time and effort it would take to develop and deploy such systems—or to expand their throw capacities—could be considerably less than one might expect, whether these systems are constructed entirely from components brought from Earth, or in part from native materials.

[3]For elaboration, see ibid.

2. Aerobraking, which uses the natural resistance of the Earth's atmosphere to slow down an entering object, may offer a way to bring packaged, nonterrestrial materials into low Earth orbits (LEO) of any inclination with equal ease from the moon or asteroids. Once in LEO, packaged materials could be provided with low mass heat shields, which would facilitate their final descent to the Earth's surface. Such shields, moreover, could be built on Earth, or constructed from materials found on the moon. Several procedures for utilizing lunar resources have already been suggested, and can be developed further with present knowledge. As familiarity with particular near-Earth asteroids (NEAs) increases, in-situ heat shield production should become even more practicable.

3. Lunar and asteroidal materials can be transformed as well into chemical propellants for use in a wide array of propulsion systems. Some could have little or no dependence on supplies from Earth for their long-term operation. This could be the case, for example, with respect to simple pressure-fed rockets that use molten aluminum and liquid oxygen.

4. Greater access to relatively large quantities of lunar and asteroidal resources might also facilitate their use in a variety of space-related ventures that are difficult and costly to perform by means of Earth-based resources and capabilities. This would include the provision of propellants and other essential supplies to spacecraft at key points in a mission; the use of the momentum created by in-falling materials to propel (and move around) space debris and facilities located in Earth orbit; and the repositioning or destruction of ballistic objects.

5. Deploying even small-scale materials industries on the moon or near-Earth asteroids would permit the development of processing facilities capable of creating power and material supply flows far greater than those which could be supplied from Earth. After two to three growth cycles, for instance, it should be possible to establish chemical propellant production facilities, capable of supplying 10,000 tons of materials on demand through cislunar space[4] and beyond. Eventually, far larger facilities are conceivable, and several possibilities already have been studied by NASA researchers.

6. In time, synergisms between Earth and space capabilities, such as the transport of materials from Earth to LEO, will help improve the cost-effectiveness of all travel from Earth to orbit and deep space. As the expense and difficulty of space operations are reduced, bases might even be built on the moon to beam large quantities of nuclear or solar derived power to major facilities positioned near Earth. Beamed power also could be used to ener-

[4]Cislunar space extends from Earth's geosynchronous orbit—that is, 22,300 miles out—to the orbit of the moon.

gize spacecraft, which then could operate more efficiently in cislunar, and perhaps even translunar,[5] space.

Rocket transportation from Earth to space now is achieved primarily by chemical rockets, which dissipate their original stores during flight. However, many approaches are being examined to decrease the costs of space missions, while increasing payloads, functional stay times, and a variety of capabilities for in-space operations. It is certainly conceivable, therefore, that turn-of-the-century space systems will take to low Earth orbit several thousand tons per year of payload at freight costs significantly less than those which exist today.

National Security Needs[6]

The Strategic Defense Initiative (SDI), frequently referred to as the "Star Wars" program, inevitably will require heavy reliance upon space-based assets (though not necessarily space-based "weapons"). This would include surveillance sensors, data processing systems, and command, control and communications (C^3) networks, to provide constant and global surveillance of the planet Earth and those space regimes where critical assets are deployed. The surveillance sensors should yield unprecedentedly large data flows, which will require rapid processing and transmission. One can envision multiple classes of sensors (e.g., infrared optics, millimeter wave and laser radar) deployed at a variety of orbital altitudes, ranging from relatively low earth orbit to geosynchronous and beyond. Survivability will be an essential requirement for space-based assets, and multiple paths may be pursued to achieve this goal (i.e., redundance, and hardening and shielding against the effects of nuclear and kinetic energy weapons).

Whatever approach is taken, the amount of materials needed to construct an effective space component to the SDI program is likely to be substantial. Hence, given the expense of transporting tons of goods to a wide range of orbits about Earth, it could prove more cost-effective to acquire most (if not all) of the materials required from the moon and NEAs. Toward that end, two alternative concepts for utilizing extra-terrestrial resources have been postulated. The first and least difficult envisions "capturing" near-Earth asteroids and then moving them to provide physical shielding for space-based assets.

The second concept calls for the actual mining, extraction, and/or processing of lunar and asteroidal materials. Possible activities range from the landing of

[5]Translunar space extends from an orbit beyond the moon, where the gravitational effect of the sun becomes greater than that of Earth, to the edge of the solar system.

[6]For further elaboration on these aspects of space policy, see Uri Ra'anan and Robert L. Pfaltzgraff, Jr., editors, *International Security Dimensions of Space* (Hamden, Conn: Archon Books, 1984).

"miners" equipped with processing equipment on selected asteroids, to the "capture" of asteroids and their transfer to lunar or space-based factories. This latter approach offers the potential to produce large amounts of "processed" materials in space as well as the "mining" of certain critical materials – such as precious metals – that could be sent down to Earth. No doubt, the development of technologies to perform such missions will revolutionize America's capability to travel through, and prosper within, cislunar and translunar space.

Over the long run, then, national security concerns, as well as space transport costs and technological advances, will strengthen the case for exploiting lunar and asteroidal materials. So will the need to develop alternative sources for critical and strategic resources on Earth that are vulnerable to supply disruption, although this concern remains of secondary importance to current materials-related initiatives in space. Yet, whatever the precise motivation, the conclusion is inescapable: The development and utilization of space resources is almost certain to emerge as an increasingly important adjunct to our overall mineral policy and program.

Mr. R. Daniel McMichael, who served as Chairman of the Forum, and Dr. Charles Perry (left to right).

Dean Theodore L. Eliot, Jr., Mr. R. Daniel McMichael and Professor Uri Ra'anan (left to right).

Mr. R. Daniel McMichael, Lt. Col. Roger Garrett, USAF, Professor Uri Ra'anan, Mr. E.F. Andrews, Professor Robert L. Pfaltzgraff, Jr., and Ms. Dorothy Nicolosi (left to right).

Mr. Richard B. Levine and Dr. William Schneider, Jr. (left to right).

Forum Session.

Mr. Richard Donnelly addressing the Forum.

Mr. Paul Krueger and Mr. R. Daniel McMichael.

Mr. E.F. Andrews and Mr. R. Daniel McMichael.

Mr. Richard Donnelly, Mr. E.F. Andrews, Mr. Paul Krueger and Mr. R. Daniel McMichael (left to right).

The Honorable James D. Santini delivering the luncheon address.

Mr. Robert Wilson, Dr. William Schneider, Jr. and Mr. R. Daniel McMichael (left to right).

Dr. John Thomas and Mr. Robert Wilson.

Private Sector Concerns and Initiatives

*by E.F. Andrews**

The history of the strategic minerals problem, the resource war, our import dependency, the vulnerability of America's lifeline—all these factors are reasonably well known. However, I do not believe that there are any actual shortages. *Access* in time of need is the problem, and, of course, the ability to process materials and to deliver them. We are beginning, moreover, to focus increasingly on the fact that, even though we might have access to resources, if we do not possess the ability to extract, process, and deliver them, such access will not suffice. Minerals in the ground are valueless. Essentially, such material is dirt until it hits a furnace or a machine of some sort. Thus, we must not forget these issues when worrying about geopolitical access, even though the latter continues to be the number one consideration. If one cannot reach the minerals, one can have all the processing equipment in the world, but it will not have much utility.

Moreover, the increase in consumption that we have witnessed since World War II in the United States, paralleling the rise in our standard of living, shows no sign of abating. This is reinforced by the technological explosion that has catapulted us into the TV World, the Computer World, the Man-on-the-Moon World, and the Jet Engine World—all during the last four decades. A whole generation has grown up taking all of these qualitative changes for granted and assuming that life without them is well-nigh impossible.

In addition, the process of decolonization in some fifty countries, primarily in Africa, has brought with it a changing relationship with the new owners of the materials in question. That process, admittedly, is approaching the end since there are not many areas left to decolonize. However, the political instability accompanying decolonization is still very much with us, and we have a long way to go before stability is likely to be achieved. Arab oil producers taught us that countries possessing essential materials are ready to form cartels and will not shy away from charging unconscionable prices, so we have to be pre-

*Mr. Andrews is Vice President for Materials and Services of Allegheny International, and a member of the National Strategic Materials and Minerals Advisory Committee.

pared to witness such behavior patterns again, if the governments in question see a suitable opportunity for repeating these policies.

The search for resources is going to expand considerably, including mining of the bottom of the sea and utilization of the opportunities opened up by our successes in space; obviously, research and development will continue at an accelerated pace, and conservation will be practiced—and should be, even in times of plenty. However, substitution and the utilization of synthetic substances may be difficult to justify for economic reasons. What inducements can one provide for those capable of producing a substitute material, when someone else still possesses many hundreds of years of reserves of the real substance, and would like to sell the latter for approximately 50 cents a pound in order to remain competitive? In such a case, the tendency is to try to find a way of dealing with the current producers, rather than investing funds to find a substitute.

Thus, there are certain constants that are likely to be with us for a considerable period of time, including increased consumption, more technological breakthroughs, but also political instability and terrorism. The art is to learn how to coexist with these factors. Private industry has taken certain initiatives, if only because of urgent need. For instance, major efforts were made to produce jet engines without an additional 400 pounds of cobalt—for the simple reason that the metal was unavailable in the necessary quantities. When forced to the wall, the private sector may take a little while to solve its problems, but it will come through in the end.

Moreover, industry has played an important role in raising public awareness of these global problems, through such institutions as Chambers of Commerce, the National Association of Manufacturers, and many others that have sponsored discussion programs on critical materials, thus carrying out an important educational task. Together with the government, the private sector has contributed greatly to research and development, to risk assessment, and to the solution of managerial and operational problems, even if these achievements do not appear on most official charts.

However, major concerns remain. A great deal still has to be done with regard to educating the American public, which does not yet realize that, without chrome and cobalt, it will not have a restaurant in which to eat, a hotel in which to stay, an automobile in which to drive, or a microwave oven in which to cook. The realization of the impact that shortages of such metals can have would do a great deal to arouse the average American—not to mention even more vital areas, such as the impact upon hospitals in which life-or-death surgery must be performed. Even the conveniences mentioned above, which once were considered luxuries, are now taken for granted; indeed, they are viewed as essentials. Public awareness of these facts can be raised, and should be, so that we may avoid policies that, for instance, could lead to a break in our supply of chrome.

What is more, it is necessary to create a sense of urgency and of selectivity in our priorities, initiating actions at a time when problems are not yet visible to the naked eye, thus making possible calm and prudent planning. It is essential to increase our data base, since we have created problems for ourselves in certain sectors because of lack of adequate information. With regard to the cobalt scare in the late 1970s, for instance, there was no real shortage; rather, there was a shortfall of information, and the result was that the price of cobalt shot up to $50 a pound, simply because we were floundering in a world of the unknown. The international data base is simply inadequate, a problem that can be resolved. It is worth recalling that in the 1972-1974 period, it appeared as if we were beset by major shortages, leading to limited allocations; but, now, it is realized that during this very period the national purchasing managers of the United States *increased* their raw materials inventories every month from mid-1972 through 1974, in terms of goods on hand. Supposedly, we were running out of every major item, and yet the gross national, physical units of inventory on hand went up each month. At the time, I was writing the economic survey report for the National Association of Purchasing Management, and recording the statistic; needless to say, I complained loudly about the discrepancies between appearance and reality.

In addition to a better data base, we must also set our priorities, going back, in particular, to the beginning of our land use policies. Where would this country be today had we adopted current land use policies at the time of the Louisiana Purchase? Fortunately, we decided, instead, to give people the land and to say "use it and grow strong and wealthy." Could we have become the America of today with today's kind of environmental priorities and land use policies? As it is, one-third of our land mass is now virtually under lock and key. We need better education about the drawbacks of the present condition.

I am concerned about the national stockpile. The private sector complains that the transactions over the years have been disruptive. The very existence of the stockpile is, indeed, disruptive. We must totally readdress the stockpile situation and move immediately to take it off the balance sheet. We will never solve the problem as long as FEMA, GSA, OMB, Defense and the Congress are involved in its day-to-day management. We must be innovative. I have recommended for years that it be set up as a separate government-owned corporation or board similiar to the Federal Reserve Board. As a profit center, it could be staffed with the skills necessary to structure complex multilateral transactions. It must be free to go in and out of the market quietly and daily. We know from experience that almost any force that operates daily in the marketplace is soon accommodated by the marketplace and is no longer disruptive. This way we rotate the stock to keep it in state of the art condition and sensitive to changing defense needs. As pointed out, I am concerned that we may not have the will to set aside the turf battle and accomplish the above results.

Americans, and Westerners in general, have not yet discovered a way of dealing with the developing countries, and, in particular, those of their leaders who are reasonably friendly to us. For instance, the quota of the stockpile of nickel was set at a rather low level after World War II, because we had a very stable source of this metal close to our border and owned by the United States government. That was in Cuba, where the nickel deposits were owned by the United States and controlled by Batista, who surely had his shortcomings but was on our side, and we set about getting rid of him and ended up by having someone worse. This is a pattern which we seem to follow time and again. We helped to get rid of the Shah and ended up with Khomeini. We helped to get rid of Somoza and ended up with the Sandinistas. Somehow we have not learned how to deal with those on our side, whether they be pleasant or not. We remove Western skills, because they are deemed to be "oppressive" or "colonialist," and the result frequently is a deterioration in health, food, productivity, and other desirable aspects of life. The question is whether those countries, and the rest of the globe, can afford to let this go on. Moreover, this deterioration tends to occur at the very time when a rise in productivity and health standards is essential, and one is then confronted with the dilemma of sitting by passively or being guilty of "interference." It is a complicated problem that should not be minimized; we have not yet achieved a resolution. In some instances, we have attempted to help with IMF money, but that, too, has been known to backfire.

An even more serious issue relates to the linkage between trade and ethical considerations. We have to find a way of continuing to trade with a country, continue to have diplomatic relations, without implying, in any way, that we agree or disagree with the internal politics of the government concerned. Doing business with someone does not mean necessarily that we agree with him. As the saying goes, "Being exploited by a capitalist state creates a condition of misery second only to that created by not being exploited...." Current cases of rampant starvation seem to underline this point.

Another factor that requires consideration is the increasing amount of the world's known reserves of strategic materials that is held by state-owned enterprises. Nearly a third of the resources of the Western world are currently in this category, plus one-half of those of the developing countries, and all of the reserves of the countries behind the Iron Curtain. A former member of the Department of the Treasury has defined the free market as having "no priorities, no values, no ends, no purpose, and making do with man as it found him." In refusing to bend to man's purpose, the free market protects us from the purpose of others, and it also liberates the enormous energy of self-interest. This factor, i.e., freedom, has made America great, along with the right to grow and to have and to own and to be and to protect and to watch your self-interest and to be rewarded for it. However, there is a large part of the world that does not follow the same rules. For instance, we have had contracts with countries which, when the price went in our favor, merely said, "*force majeure*," and there was no law in the world that could hold them to that contract; on the other hand, when the price went in their favor, they could take us to court.

There is the additional problem of the influence of interests that give priority to considerations other than "what is the best possible mode of production?" For instance, when the objective is simply employment, or the employment of a particular favored group, the result may be the production, or overproduction, of materials other than the ones most urgently required. While speaking about the "level playing field" as the only objective of the private sector, the truth is that industry has run to the government, time and again, asking for a quota, a tariff, or a subsidy. In other words, since "we can't whip them, we are joining them." One of these days, we may see copper on the endangered species list of American industries. Can we afford that? What is the answer? Nationalization? And the question remains how to come to grips with the problem of two great incompatible systems in the world marketplace.

All of these concerns must be addressed, and we are not very far along in our attempts to do so.

Congress and Strategic Minerals Policy

*by The Honorable James D. Santini**

The strategic minerals problem can be reduced to three negative scenarios: supply interruption and price manipulation (à la oil in 1973), decimation of our nation's processing industry, and reduced access to our own domestic minerals. The relevant Executive agencies have attempted to address parts of these problems; and, as past chairman of the House Interior Committee's Mines and Mining Subcommittee, I endeavored to pursue possible Executive agency and Legislative responses. My interest in matters mineral is longstanding. Minerals and mineral exploration, production, and processing have played a vital role in the economic health of my silver state, Nevada, although, given the fact that 80 percent of its surface has been turned into public land and another 13 percent is in the hands of local government, Nevada was forced, in 1932, to become also a gold state—by adopting legalized gaming.

It is worth noting, however, that Nevada became a state in 1864, because it was felt that its mines would prove particularly useful in support of the Union during the Civil War. When I came to Congress, it was natural for me to seek a seat on the Mines and Mining Subcommittee of the Interior Committee. Unfortunately, I encountered in Congress and elsewhere a deep prejudice aroused by the very words "mines" or "mining," which conjure up pictures of ravaged hillsides in Appalachia. The assertion that mining activities have displaced no more than .02 percent of public land seemed to fall on deaf ears. Moreover, it was difficult to discuss substantial resource issues with colleagues who had little information on, or interest in, tungsten, molybdenum, or cobalt, an attitude that merely reflected the attitude of their constituents. Consequently, the debate was frequently simplified to such comparisons as: "Do you love national forests, or do you love people who rip up hillsides?"

---

*Mr. Santini, former Congressman from Nevada, was Chairman of the House Subcommittee on Mines and Mining from 1974 to 1982. He is currently a senior partner in the Washington law firm of Bible, Santini, Hoy & Miller as well as a member of the National Strategic Materials and Minerals Program Advisory Committee.

In the first instance, we are confronting a problem of education. With the exception of states like Nevada, minerals do not have major constituencies. One might say almost that "rocks don't vote." Initially, I decided to take the resource issues to the Executive branch. A bipartisan group of legislators went with me in June 1977 to meet with President Carter and his Secretaries of Energy and Interior, with representatives of the Departments of State and the Treasury also in attendance. I was allotted seventeen minutes for a presentation on the subject of our mineral dependency and declining processing industry. At the conclusion, the President, who described the situation as "alarming," instituted a "National Non-Fuel Policy Review." I appreciated that response, just as he may have appreciated the fact that we did not take more than seventeen minutes of his time.

We felt that we had reached a turning point. Indeed, the Executive branch expended over $3 million, involved 14 government agencies, and thousands of man-hours on the non-fuel mineral study. Two years later, in the summer of 1979, after an extensive expenditure of time, money, and energy, there was a "preliminary review" of this study. After going through the excisions, revisions, and amendments submitted by 14 government agencies, the initial report was uniformly condemned by such diverse organizations as the Sierra Club and the American Mining Congress. In other words, the Report authors had managed to disappoint every institution that had an interest in the subject. Thus, the "preliminary study" terminated the whole effort.

Consequently, I next turned to the Congress to try to obtain problem-solving legislation and, thanks to the initiatives of the Science and Technology Committee in cooperation with the Interior Committee, we were able to produce the National Minerals Policy Research and Development Act of 1980. This legislation did not exactly revolutionize national perceptions or responses to minerals problems, but it did provide a legislative starting point for actions by the Executive. For one, the Executive agencies were mandated to report annually to the Congress on matters concerning materials and minerals, so as to indicate whether Congressional action was warranted. During the first year, only one agency, Commerce, made any effort to be responsive to this Act. The General Accounting Office felt impelled to castigate the Administration in order to secure its compliance with the requirements of the law.

I have thought that it is of vital importance for a national government concerned with American productivity to note the decline of the minerals processing industry—indeed, its calamitous downturn. Whether we look at Anaconda, ASARCO, Kennecott, Lead Zinc facilities, or the steel industry, our national government should be responding to the problem. The censure of the Executive branch by the General Accounting Office concerning the lack of response and the consequent absence of Congressional follow-up did produce a report from the Reagan Administration in April 1982. However, again the General Accounting Office was compelled to report that the 1982 effort did not respond to the law, since it failed to bring together the necessary basic facts and figures, and failed also to advise both the Congress and the Executive agencies what should be done about a declining processing industry, the potential vulnerabilities created by the dependency on foreign sources, and the implications for America's national security and commerce.

Consequently, the Congress again proceeded to take some legislative initiatives, including the authorization of $80 million in the fall of 1982 for the purchase of domestic copper for our national stockpile. However, the Executive agencies simply did not follow up so that, during this depressed period for the U.S. copper industry, no action was taken.

The problem is basically that the minerals industry is victimized by *ad hoc* decision-making within each department or at the Cabinet level. It all depends on which government agency is making the decision. If it happens to be Treasury, no one wants to spend a dime. If it happens to be the Department of State, there is no interest in resources when there are international accords and relations to worry about. Resources rate a very low priority. Even in our diplomatic missions in the countries of Southern Africa, the source of most of the strategic minerals needed by the Free World, the United States has not posted a sufficient number of minerals specialists to look after our national interest in this critical field. This oversight is indicative of the kind of focus and emphasis provided by our national government. As my friends in the Department of State say, "Hey, look, we have enough headaches already, with limited personnel and budget, having to deal with enormous global problems, without having to worry about obscure issues like manganese dependency."

Eventually, in 1984, with the House Science and Technology Committee leading the way, we were successful in establishing a Materials Council, situated at the White House level, to rectify, at least in part, the *ad hoc* decision-making on the minerals and materials issues. Unfortunately, as of this writing, the Administration has not implemented the Council.

Of all the mineral resource problems and frustrations, the copper issue is one of the most vexing. In July 1984, the U.S. International Trade Commission made the determination that either a tariff or quota should be imposed on copper coming into the United States. This came from an agency of the Executive arm of the government that is supposed to review facts and figures and make a decision. The tariff quota recommendation was then sent up the decision-making ladder to the White House decision-maker, Mr. William Brock, and the Council of Economic Advisors (who have other sizable economic worries on their agenda). I doubt whether they have experience or knowledge pertaining to the copper industry, and the copper industry has no economic impact on their principal concerns, international or domestic. The Council evaluates issues from the standpoint of theoretical (and real) disruptions of their "free trade" commitment. Consequently, while the Council endorsed a 25 percent quota on the import of foreign cars, it rejected consideration of any quota or tariff relief to a copper industry in disastrous decline. The problem is that decisions are made at the highest levels of government without any direct factual input from the best-informed minds in our country—such as many of the contributors to this volume. This is exasperating, because the balance of give-and-take has gone and the information input is one-sided or erroneous.

As a result, there are likely to be more initiatives by the Congress. In 1984, for example, Senator DeConcini proposed that not more than 300,000 metric tons of copper should be allowed to come in during a three-year period. When it is noted that in 1983 alone, over 500,000 metric tons of copper were imported, one can appreciate the importance of that legislative initiative. Senator DeConcini also suggested that the International Monetary Fund's compensatory financing facilities be prohibited from giving any loans or financial assistance, if it were determined that such a loan would increase the capacity of an industry that was manufacturing a product already in world surplus. This constituted an understandable expression of frustration by a Senator who represents a state which is experiencing the radical decline of the American copper industry.

In the Paley Commission era of the early 1950s, the United States had in place a suitable response to a military or war-generated resource or material need. There was a genuine effort to try to institute a plan, a policy, that truly reflected the sense of urgency to take action, with both national and global implications, to stabilize America's ability to manufacture a product and to obtain the ingredients to make that product. Today, the fact is that airplane engines require cobalt as an indispensable manufacturing component and we obtain cobalt from a country beset by economic, social, and political instability—namely, Zaire—and from a less-than-friendly government next door, Zambia. Over 50 percent of the Free World's cobalt comes from these two countries. Nobody has dealt realistically with the scenario of a cutoff from that part of the globe. In his paper, Mr. McMichael has described the effects of a chromite cutoff from Zimbabwe and South Africa, resulting in the unemployment of over a million Americans. These are real jobs and real products, and the threat of a cutoff poses a very real problem. Meanwhile, misunderstanding and lack of interest persist in both the Legislative branch, where I endeavored to implement a solution, and the Executive branch, in which Paul Krueger struggles as one of the few disciples of light. It is frustrating, because a plan or policy should be in place. That is inherent in any kind of problem-solving; you cannot address the ramifications of the problem—whether it concerns copper, cobalt, manganese, or chromium—until you have some sort of plan or scenario to deal with it.

We have a national commission now—the National Strategic Materials and Minerals Program Advisory Committee—which constitutes an Executive response to a Legislative mandate to do something about mineral problems. Will this 25-member commission do something about it? During its two-year term of existence, it has an opportunity to bring back specific recommendations for Executive action. First, we must have a complete assessment of our domestic resource capacity, together with an inventory of the resources present on public land in the United States. We do not have the total picture, despite the preliminary regional fact-gathering conducted by the Bureau of Mines and the U.S. Geological Survey. We do not know how many acres of public land in the United States are off limits to exploration and mining. The Congressional Budget Office estimated last year that half of the public lands (one-third of the land in America on which most of the mineral potential exists) is off limits. The truth of the matter is that neither the Department of the Interior nor the U.S. Forest Service—nor, for that matter, any other agency of the government—knows how much of the public land is off limits to exploration or minerals recovery.

The Advisory Committee, headed by Admiral William Mott, USN (Ret.), and reporting to the Secretary of the Interior, can make aggressive recommendations, but first we have to find out what is on and off limits, and what we can or cannot recover. Our Advisory Committee can do that. We have had former Interior Secretary William P. Clark's enthusiastic endorsement because he appreciated the national security implications. However, there will be no magic wands or formulas that will transform the citizens and politicians of this nation into an enlightened aggregate of enthusiasts about tungsten. All that we can do is, bit by bit and piece by piece, start making some important changes at the national government level.

A View from the House of Representatives

*by Paul C. Maxwell**

Congressional interest in strategic minerals traces back more than half a century. Critical commodities lists were created during and after World War I; a fledgling national stockpile, presaging World War II, was created in the 1930s; the Cold War and Korea led to a broadened defense stockpile and introduction of the Defense Production Act of 1950. During the past thirty years, strategic materials have been the focus of numerous studies, committees, and commissions—perhaps the best known being the Paley Commission established in l951. Most of these focused on the problems related to major national emergencies leading to a declaration of war.

However, it was only with the crises created by the oil embargo of 1973 that a fuller realization of the nation's minerals/materials vulnerability during peacetime became apparent. Oil was only one of many items for which the United States could find itself held hostage to a supply interruption or cutoff. The cobalt crisis of 1978, for instance, triggered by guerrilla uprisings in the Shaba province of Zaire, underscored our materials vulnerability. The resulting panic in commercial markets drove prices for cobalt from $6 per pound to more than $50 per pound and caused shortages lasting over 18 months. Concern mounted for other commodities, with chromium, manganese, and platinum—all imported at greater than 90 percent levels and all largely supplied and produced by Southern African states—receiving the most attention.

Congress responded by enacting the National Materials and Minerals Policy, Research and Development Act of 1980 (P.L. 96-479). While attempting to address more comprehensive concerns, the strategic implications of critical materials were a major thrust of the legislation, calling for the Department of Defense to report to Congress and the President on the impact of strategic materials on defense programs. This assessment was intended to be incorporated in a broader policy framework to be developed through the White

---

*Dr. Maxwell is Science Consultant for the House Committee on Science and Technology, U.S. Congress.

House and presented to the Congress as a Presidential program plan for addressing national materials problems.

With the advent of the first term of the Reagan Administration in 1981, formation of the cabinet-level Natural Resources and Environment Council, a spurt of activity by the Departments of Defense, Commerce, and Interior, as well as numerous hearings, articles, and outside reports on critical materials, led many to believe that this decades-old problem was at last moving toward resolution. The Congressionally mandated program plan appeared to take at least some of the necessary steps in outlining Federal action in this area. The plan, noting the strategic importance of materials, placed heavy emphasis on domestic production of minerals and public lands management, as well as pointing to the importance of materials research and development. The Cabinet Council was charged with responsibility for attaining the goals of the proposed plan, though no monies were targeted for any part of it.

Unfortunately, the high expectations generated in 1981 were quickly frustrated by the reality of little or no perceptible results. The Cabinet Council, while potentially useful as the coordinating and implementing mechanism required by P.L. 96-479, was not effective. The Council met only infrequently on materials issues and then only on an *ad hoc* basis. The disbandment of the various interagency working committees left a commitment to materials policy without permanence or clearly defined lines of communication with the rest of the Federal government. Dominance of the Council's activities by the Secretary and Department of the Interior placed undue emphasis on questions of importance to that agency, while ignoring much of the needs of the rest of the government, including defense.

The argument was proffered, for instance, that a reduction of three tons of a given mineral in the Defense stockpile could be achieved through an increase of one ton in domestic production. This position failed to note, however, that the Department of Defense needs *products* made of materials, and not necessarily the raw material itself. Thus, the question of industrial processing and use of advanced synthetic materials is more important to security needs than simply stockpiling unprocessed raw minerals. Interior also neglected to account for the fact that much of the nation's domestically available minerals *do not* include the most sought after critical minerals—i.e., chromium, cobalt, manganese, platinum and others. Opening up wilderness areas and other public lands in the geographical United States is not likely to provide these essential minerals, since Mother Nature has not placed them in sufficient quantities within our borders.

The Department of Defense, as well as Commerce, did move forward more or less independently of the Interior-dominated Council. Their own analyses and studies suggested that the processing of advanced, as well as basic, materials would be of critical importance to maintaining the strategic strength of the country. In carrying out Defense's mandated study from the 1980 Act, both the Defense Production Act of 1950 and the Strategic Stockpiling Act were quickly recognized as means to address materials-related national security needs. This assessment, while acknowledging the problem of import dependence, emphasized the nation's critical reliance on its industrial capacity to convert raw materials into defense products, the need to provide for quality in the stockpile, and the importance of continuing research and development of advanced materials and processes.

This emphasis leading to an expanded role for the Federal government in supporting our industrial base through the Defense Production Act quickly ran into difficulty with a cost-conscious Office of Management and Budget (OMB). The mandated report, as a result, never reached the Congress. Subsequently, despite attempts by Congress to strengthen the legislation, the Defense Production Act was extended with inclusion of only minor support for the materials production industry.

Other studies, assessments, and even programs met with the equal fate of being ignored, or of offering little of value to the policymakers high within the Executive Branch. This was true of the assessments made by the Department of Commerce on strategic materials in critical industrial segments (aerospace and steel), as well as programs for critical materials substitution within NASA, DOE and DOD. For instance, a small research program entitled COSAM (Conservation of Strategic Aerospace Materials) flourished briefly in 1982 and 1983 and then was dropped for lack of continued interest. The Manufacturing Technology Program within DOD, while continuing, has never achieved the expected level of support envisioned at its inception.

The problem of maintaining interest in materials policy on the part of the Administration, as well as Congressional leaders, can be ascribed in part to confusion as to the character of mineral supply vulnerabilities. Early predictions of a major crisis in strategic materials on the order of another international oil embargo have not been fulfilled. The possibility of strong cartel action or major political uprisings leading to a supply interruption appears to be unlikely. The problems resulting from import dependence are more insidious and long term. More likely is the gradual erosion of the U.S. industrial base in the materials sector, an erosion already moving alarmingly ahead.

The steel industry is one case in point. The recent economic recession, while striking hard at all of the basic materials industries, has left steel especially devastated. The industry is now operating at only 40 percent of its capacity with more than 50 percent of its work force unemployed. While some plants have re-opened and some jobs have been restored, the recovery has been extremely slow. No one is predicting that the industry will return to anywhere near its former operating capacity. While mergers appear to be one means for salvaging marginal operations, the costs are extremely high. For instance, the merger of Jones/Laughlin with Republic Steel last year created LTV Steel, second in size only to the U.S. Steel Corporation. However, the merger, together with other resource changes, required over $800 million in cost reductions to keep the new company alive.

Competition from abroad is one major concern. European steel companies have been subsidized by more than $30 billion from 1976 to 1981. As a result, foreign imports have risen to over one-third of the U.S. steel market, and are anticipated to exceed 40 percent by the end of the decade. In a one year period during 1982-1983, the U.S. industry lost over $6 billion.

A similar or even worse situation exists in America's ferro-alloy and non-ferrous mineral industries. We no longer have sufficient facilities to produce ferro-chromium or ferro-manganese, with foreign supplies of these two commodities accounting for more than 75 percent of the U.S. market in 1983 (over 90 percent for high carbon ferro-chromium/manganese). While imports now account for only 29 percent of the ferro-silicon market, the Soviet Union recently began to sell to U.S. consumers, threatening even this vital industry sector.

Reasons for these continuing declines in industrial capacity cannot be laid solely at the doorsteps of an international conspiracy, either economically or politically instigated. Rather, it appears to be due in part to a number of complex factors. One is the displacement of the older "smoke-stack" industries by newer high technology, "white glove" industries. While seemingly more attractive, the high technology industries will certainly be insufficient to meet our nation's basic defense needs in the event of a national emergency.

Another reason for U.S. industrial weakness is the shortage of investment capital to promote greater use of existing advanced technology. The looming federal deficit projected over the next decade does not make resolution of this problem in the open market likely. Yet another reason offered for current difficulties is what is often called the "Harvard MBA mentality." Management and investors concerned only with short-term gains have lost sight of long-term investment needs, which are essential to maintain international competitiveness. Central to these and other problem areas has been the general lack of clear federal guidance regarding industrial policy.

The prospects for the advanced, high technology materials industries, moreover, are not much better than those for the basic industries. Again, the problem is not solely import vulnerability, but rather our apparent inability to maintain a technological edge in an extremely competitive international economic environment. Japan, Western Europe, and others are *applying* advanced material concepts to such major high technology industries as transportation, communication, and information.

Japan, for instance, has been embarked since 1980 on a major research program in advanced, fine ceramics, both structural and electronic. Over 100 companies are presently involved in a program designed specifically, and acknowledged candidly, by Japan to "leapfrog" U.S. capability in this area. Roughly $1 billion over 10 years has been dedicated to this task, with tangible results already forthcoming in such diverse products as scissors, seals, pens, and mill rolls. In testimony in 1984 before the House Science and Technology Committee, industry representatives provided evidence that Japan now has in prototype production an advanced ceramic gas turbine car, two to three years in advance of U.S. efforts. The Japanese program is geared to produce a final, commercial product for profit, whereas federally supported efforts in the United States are aimed at providing a technology, but not necessarily one that will move toward early commercialization. We will create one or two prototype models, but with little interest by industry to move much beyond that.

Unfortunately, our efforts in advanced ceramics are paralleled in other areas such as polymeric-based composites and advanced alloy systems. France, the Federal Republic of Germany, Great Britain, and others in the Common Market appear to have taken their cue from Japan and have initiated major advanced materials programs involving extensive international and inter-regional collaboration.

This problem has been complicated by recent, well-intentioned attempts to control more fully the export of advanced technology. Warnings by Admiral B.R. Inman that there existed a "hemorrhage of U.S. technology to the Soviets" has led to efforts by the Reagan Administration to stop such a "hemorrhage" through stricter interpretations of export control regulations. Advanced materials are central to many of the technologies considered sensitive. Draft guidelines to be used in establishing regulations for export control, while intending to safeguard U.S. technology developments with military utility, are viewed by many scientists as potentially severe constraints on research and information exchange.

Industry leaders, moreover, believe that the guidelines may severely encumber, if not actually prevent, effective U.S. competition in the international marketplace. For even if the United States is not in the premier position for certain areas of research, strict export controls could be imposed nevertheless, due to perceptions "that ongoing U.S. research is expected to yield near-term results of extreme strategic interest." Definition of technology export, it is worth noting, includes conversations with foreign graduate students, presentation of relevant research information at foreign technical or scientific conferences, and other activities common to normal research information exchange. Obviously a dilemma exists when the Department of Energy works to promote technology transfer of its advanced materials programs (such as the ceramic gas turbine), and the Department of Defense moves in the opposite direction with the same material.

The Congress, in reviewing materials policy over the past four years, recently passed legislation which hopefully will help in promoting solutions to many of these problems. The National Critical Materials Act of 1984 (P.L. 98-373) was signed into law by the President on July 31, 1984. This legislation, as a follow-up to the 1980 Materials Policy Act, establishes a three-member National Critical Materials Council, under and reporting to the Executive Office of the President. Congress sees the Council as the focal point for addressing *all* materials issues, responsible for advising the President and making recommendations to the Congress. The Council will make recommendations in assigning responsibilities and in helping to coordinate Federal materials-related policies, programs, and research activities. The Council will also review and assess the various programs and activities of the government in accordance with the direction and policy of the National Materials and Minerals Policy, Research and Development Act of 1980 (P.L. 96-479).

The Council is expected to oversee the Federal materials R&D policies and programs and, in particular, to establish a national Federal program plan for advanced materials research and development. In addition, the Office of Management and Budget will be required to consider materials research, development, and technology funding requests as a single, multi-agency request in each fiscal year and indicate its adherence to the materials program plan. Other actions are also to be taken by the Council to promote innovation and productivity in the basic and advanced materials industries. These include consideration of generic centers for industrial technology, data/information systems, and other actions.

The agenda for the new Council most likely will focus on the following sets of issues: (1) domestic minerals production, public lands, and environmental policy; (2) international policies related to minerals vulnerability, as well as U.S. competitiveness in advanced materials technology; (3) Defense Department needs, including advanced materials and stockpiling management or policy; (4) industrial innovation in both advanced as well as basic materials industries (e.g., aerospace and steel); and (5) materials research, technology, and engineering, including a national program in advanced materials R&D. Obviously these areas are interrelated. However, the Council represents the first opportunity for considering all of these matters in a cohesive and focused manner.

In summary, it seems clear that the question of strategic materials involves a complex set of issues. While our critical materials import vulnerability cannot be ignored, we must be aware of, and take the necessary steps to address, the problems of erosion in domestic industries and materials technologies. We are, perhaps, our own worst enemies, if we ignore the obvious decline of our essential basic industries, while failing to promote advanced materials technologies. Failure to remain competitive in a harsh international economic environment leaves us vulnerable to our allies, as well as our foes. Congress believes that the Federal government has a primary role in identifying clear, national objectives regarding strategic materials, and in taking the necessary steps to reach those objectives. The creation of the National Critical Materials Council under the President provides the Executive branch with a new mechanism to carry out this responsibility. Clearly, the value of the Council will be dependent on the quality of people named to the Council and its support staff, as well as the manner in which its advice is received by the President. Congress can be expected to follow closely the Council's activities as it works to address continuing strategic materials problems.

A View From the U.S. Senate

*by Robert L. Terrell**

While some interesting points have been made regarding the National Defense Stockpile, and about the importance of establishing a better balance between mineral and material supplies from foreign and domestic sources, there is need for elaboration. I agree with recent changes in our stockpile policy, which, in contrast to past policies, support the full spectrum of potential conflict—i.e., stockpile goals adequate to undergird at least a three-year war effort. Following World War II, stockpiling objectives were based on contingencies that envisaged five years of total war. Then, in 1958, stockpile objectives were changed to reflect a three-year war scenario, and in 1973 they were lowered further to reflect a one-year war. Since 1976, however, the stockpile objectives reflect, once again, a three-year war scenario.

It is important to note that each time the war scenario was reduced, stockpile goals were lowered, and, as a result, many stockpiled materials were declared to be in excess and were disposed of in the market place. Previous changes in stockpile goals were at the discretion of the President. The current stockpile goals were statutorily established with the enactment of the Strategic and Critical Materials Stockpiling Revision Act of 1979. The point is that many of the materials in the stockpile that were declared to be in excess over the last three decades or so were subsequently determined to be strategic and critical at a later date. For example, the stockpile goal for lead in 1944 was 1,100,000 short tons, and in 1963 it was reduced to zero. Today, however, the goal for lead is 1,100,000 short tons, once again, and in most instances the taxpayer has borne the brunt of these disposals and subsequent repurchases.

There is also an ongoing disagreement today with regard to 137 million troy ounces of silver in the stockpile, which the Administration considers to be in

---

*Mr. Terrell is a Professional Staff Member working with the Senate Committee on Energy and Natural Resources.

excess of our defense needs, but which Senator McClure[1] believes to be necessary. Part of this disagreement revolves around alternatives for disposing of the silver. If silver is in excess and must be disposed of, Senator McClure prefers that it be released in the form of coins, rather than by means of an auction of excess stocks run by the General Services Administration—a procedure which could severely disrupt the world market and adversely affect domestic silver producers. Who is to say, moreover, that with changing technologies, we might not find ourselves in need of silver in the stockpile later on?

The point also has been made that the stockpile is being reviewed as to the quality of materials on hand. Clearly, that is important and should be done. However, I believe that a review should also be undertaken with respect to the location of stockpiled materials in relation to those industries that actually would be called upon to process released stocks in wartime. As many are aware, most stockpile sites were selected over a quarter of a century ago, and were situated in close proximity to key processing industries. A large number of these industries, however, have relocated over the years, so there is considerable uncertainty as to whether stocks could be transferred in a timely manner to the appropriate industries in the event of a national emergency.

In discussions of the history of the national defense stockpile, the Korean War era generally is identified as a time when the United States acquired a tremendous amount of stocks. Moreover, attention often is drawn to the government loans and purchase guarantees provided under Title III of the Defense Production Act (DPA) of 1950, which helped to bring on stream additional domestic mineral and material resources as a supplement to—and a source for—the stockpile. Yet, in addition to DPA initiatives, there was considerable reliance in the past on the barter trade authority contained in the Agricultural Trade Development and Assistance Act of 1954, which was used to create a "supplemental stockpile" of strategic and critical materials. I mention the 1954 Act because the issue of barter resurfaced in the 98th Congress. Proponents of barter recommended, for example, that we exchange surplus wheat for imported oil for the strategic petroleum reserve—a bushel for a barrel. While this seemed reasonable, the Department of Agriculture's Payment-in-Kind (PIK) program prevented such transactions. Nevertheless, there has been a successful use of barter by the Reagan Administration involving the exchange of surplus U.S. dairy products for Jamaican bauxite to be placed in stockpile.

In any case, it should be noted that barter trade based on the 1954 Act has been extremely fruitful. By the end of 1961, the strategic and critical materials

---

[1]Senator James A. McClure of Idaho is Chairman of the Senate Committee on Energy and Natural Resources, which deals with a host of mineral and material policy issues, including the stockpile.

acquired through bartering, and subsequently transferable to the stockpile, were valued at $223,243,000. The total value of materials acquired for the "supplemental stockpile" was $1,103,506,000, of which $961,920,000 worth had been acquired through bartering. The remainder was acquired through direct purchases using foreign currencies gained through the sale of agricultural goods overseas.

Use of the term "supplemental stockpile," by the way, results from the fact that strategic and critical materials acquired through barter had to be maintained by the Commodity Credit Corporation in the Department of Agriculture until such time as they could be sold to the appropriate federal agency. While exercise of the barter authority contained in the 1954 Act was very successful, it was stringently opposed by the Commodity Credit Corporation. Essentially, the Corporation's opposition was based upon the fact that items acquired through barter could not be sold in an expeditious manner to the appropriate federal agency. Yet, the Committee on Agriculture in the House of Representatives concluded that the Commodity Credit Corporation's funds and assets could be better protected by exchanging, when the opportunity was offered, some of its costly-to-store agricultural supplies for nondeteriorating, easily stored strategic materials, even though these may have to be held for some time as the property of the Corporation. Indeed, to refuse to make such exchanges simply because no government agency was in a position at the moment to buy the strategic materials from the Commodity Credit Corporation, was to negate the very reason for barter—which is an exchange of materials for materials, when money with which to purchase needed commodities is unavailable or is considered less useful than the ready availability of such commodities.

The cost advantages of stockpiling through the barter of goods have been striking. It was estimated in the mid-1950s, for example, that the storage cost for a ton of wheat was about $5 a year, compared to one-fifth of a cent for a ton of ore. I believe that the use of barter today is as valid as it was then, and for the same reasons. Unfortunately, this view is not shared by a majority in Congress, or in the Administration. Indeed, efforts directed toward industrial preparedness and mobilization are severely hampered by the absence of a clear and pending threat similar to those which existed during the Korean and Vietnam wars.

Lately, there has been much discussion of the efforts made by Congressman Fuqua and his staff at the House Science and Technology Committee to pass legislation creating a National Materials Council in the Executive Office of the President that, as one recent bill suggests, "hopefully will rectify at least part of the *ad hoc* breakdown in communications about an intelligent overall approach to problem solving of mineral materials issues." These efforts led to the passage in July of Public Law 98-373, entitled "Arctic Research and Policy Act of 1984," Title II of which makes provision for a "National Materials Council." It remains to be seen, however, whether this Council actually will rectify the "*ad hoc* breakdown in communications" concerning "an intelligent overall approach to problem solving." As a matter of fact, I view this new layer of bureaucracy as creating an overlapping mechanism which, because of its expansive role, still will have to rely upon existing federal agencies for information, advice, and analysis. This will create additional burdens for the federal agencies involved, and, in many instances, will result in duplication of effort.

Proponents of this measure had one basic objective in mind—the establishment of a "clearing house" or a "one-stop shopping" office for minerals policy issues and the allocation of research funds. The practical result, however, may be to confuse and complicate the policy process. Those existing federal agencies responsible for mineral and materials technology and policy input are knowledgeable in their areas of responsibility, and are cognizant of industry's problems. Imposing another level of bureaucracy above that which exists already is unlikely to enhance the knowledge of, or the emphasis given to, minerals and materials issues by this or any other Administration.

Moreover, the new Council might interfere with the operation of our free marketplace. In some respects, the creation of such a Council constitutes the beginnings of a formal national industrial policymaking group. Yet, the need for such a group has not been established, and the whole concept is likely to have far-reaching implications for the American economy. Even if this were not the case, and there was a real need for the White House Council, Public Law 98-373 would not satisfy that need. Indeed, after close examination of the purposes and objectives of the law, one clearly can see serious redundancies with statutes already on the books. As a matter of fact, a comparison of P.L. 98-373 and four other existing statutes would show that provisions contained in the new bill are virtually identical with legislation previously enacted; and since none of these pre-existing statutes were modified or repealed by the new legislation, federal agencies set up under earlier laws will continue to operate irrespective of the enactment of P.L. 98-373. This, then, may be a case where Congress is legislating through tunnel vision, not looking back at previous law.

There are problems also with the perceived need to inventory the resources on public lands, and to determine how many acres have been placed off limits to exploration and mining. While I agree that it is useful to ascertain how much public land is off limits to mineral exploration and mining, I disagree with the idea that the government should inventory public lands in an effort to document mineral resources. I should point out that Section 204 (1) of the

Federal Land Policy and Management Act of 1976 already requires the Secretary of the Interior to review all public lands, including those in the National Forest System, which are closed to exploration under the Mining Law of 1872, and to leasing under the Mineral Leasing Act of 1920. This review is to be completed by 1991.

As for suggestions that the Federal government attempt to inventory public lands to determine their mineral resource content, I am not sure that this effort is necessary. One only has to recall the problems encountered by the U.S. Bureau of Mines and the U.S. Geological Survey in their attempts to determine what mineral values may exist in lands proposed as components of the National Wilderness Preservation System. Such attempts are cursory at best, and ultimately disappointing for two reasons. First, these agencies lack the incentive that private industry has to explore for the existence of mineral resources. Second, and more important, their efforts are often constrained by inadequate funding from Congress. Information derived from the cursory examinations performed by the U.S. Bureau of Mines and the U.S. Geological Survey in prospective wilderness candidate areas, nevertheless,is often taken as *prima facie* evidence that valuable mineral resources do or do not exist. If these types of examinations had been conducted in the past in Texas, Oklahoma, and Louisiana by federal bureaucrats, it is quite likely that oil and gas would not have been discovered. The only realistic way to find out what mineral resources might exist on the nation's public lands is to provide access to those lands to private industry, whose business it is to explore, develop, and produce such resources.

On the issue of U.S. import vulnerability, I tend to agree with those who argue that we need to strengthen our domestic industries and render them more competitive. In the 1970s, we had the economic globalists advocating interdependence, not from the perspective of free trade, but as an element of national policy. They argued that U.S. dependence on imports should not be viewed as vulnerability, but as a step toward a safer world attainable through ever expanding interdependence. This same group accepted and supported the demands of the Group of 77 for guaranteed financial assistance and technology transfers to increase Third World processing capabilities, which, by diminishing American processing capabilities, could decrease the ability of the United States to meet essential supply needs during a national security crisis.

Even officials in the U.S. Bureau of Mines, appointed by the Carter Administration, have argued that there is little need to worry about our increasing dependence on foreign suppliers, since such concern represents an oversimplification of the dependence-vulnerability equation. This was a significant departure from the Bureau's long-held and more fundamental position in support of a strong and reliable domestic minerals industry. I would recall, moreover, that when the British House of Lords was examining the strategic stockpile question, its expert witness considered vulnerability to exist when the number of suppliers of a commodity numbered two or less. We take it for granted that the United States is a reliable supplier of those few minerals we export, but both France and Japan have categorized molybdenum—whose production and reserve base is dominated by the United States—as a priority commodity on their stockpile list.

As to current debates over the strategic classification of copper, lead, and zinc, it is worth noting that all three do, indeed, fall within the definition of strategic and critical. Furthermore, all three are short of stockpile goals by a combined 2.5 million tons. We lose sight of the importance of these large tonnage base metals, simply because in the past we have been able to supply most, if not all, of our requirements from domestic sources. Yet, we close down our smelters and refineries without so much as a backward glance, if their continued operation would conflict with environmental priorities. Moreover, we have given no priority to the location and development of new base metal deposits, and seem to have gone out of our way to thwart their development on the nation's public lands. The overriding management priority for our public lands is dedicated to their non-mineral use.

On the issue of protectionism versus free trade and our ability to stay competitive, I do not believe that the domestic minerals industry wants trade protectionism. The industry, however, does want to see some concern for *fair* trade on the part of the American government, since the United States has been notoriously slow in taking up unfair trade issues with our foreign competitors. In the recent case of copper, for example, it is my view that the industry, rather than seeking relief through tariffs or quotas or even orderly marketing arrangements, would much rather have seen our government quietly negotiate voluntary production constraints with those government-owned foreign producers, who, over the last three years, have continued copper production at capacity levels, despite the existence of a world copper market that is severely out of balance. We say that our private producers operating under a profit system must take their signals from the market, yet we overlook foreign producers that produce for purposes other than business profit. When commodity prices drop and export earnings shrink, the fallback position of foreign producers is to apply for a compensatory financing facility loan from the International Monetary Fund (IMF). Thus, while we apply the rigors of the market to domestic U.S. producers, we tend to overlook market requirements when we support compensatory loans to foreign producers.

Some comments are in order on behalf of the Reagan Administration's efforts to comply with the Mining and Materials Policy, Research and Development

Act of 1980. We often hear demands made in support of minerals policy to the effect that somebody must *do* something. However, in most instances, if we wish to accomplish an objective favorable to minerals, we must *undo* barriers to the attainment of that objective, even if it is only a change in the priority of goals. The point is that in the name of free trade, a better environment, safer work places, national health, an equitable tax system, or whatever, we have often hitched our general policy goals to specific issues that have great public appeal, but usually wind up as legislative mandates that are expanded upon, and embossed by, single-minded bureaucrats. A balanced policy then becomes well-nigh impossible.

For example, when Congress passed the 1980 Act, as well as the 1970 Mining and Minerals Policy Act, it had no solution to the general problems it saw developing as a result of our increasing inability to produce domestic minerals. Nor did it understand the larger implications of unnecessary dependence on offshore mineral suppliers. After numerous hearings and debates, with a lot of postulating thrown in, Congressional committees ended up with two laws that are merely statements of policy. They contain no specific instructions on how to accomplish the goals sought. As a case in point, the 1980 Act primarily is a requirement to do more studies which, it is assumed, somehow will come up with the answers. But we have had more than enough studies already, and simply adding to what we have will not provide solutions. So for me to hear that only the Commerce Department complied with the 1980 Act in putting out two more reports, which duplicated what was already on record, seems to "gild the lily." Congress, itself unable to assign minerals policy goals on a national list of priorities, left finding solutions up to the Executive; but in providing more specific directions on other laws, it also made solutions almost impossible. Without being too pessimistic, it should be noted that, until a real national interest in minerals security evolves, other national aims and priorities will probably predominate.

What has troubled me are the suggestions, emanating primarily from the Washington offices of the mining industry, to the effect that President Reagan's April 5, 1982 National Materials and Minerals Program Plan and Report to Congress was not a specific enough response to the goals of the 1980 Act. In effect, these industry recommendations would have the Administration assign priorities that even Congress was unable to provide in all of its deliberations leading up to passage of the 1980 Act. I happen to think that the President's policy statement was a strong document, the first in many years to restore some balance in national thinking with respect to U.S. mineral interests. As a policy statement, it was not meant to be specific. Moreover, considering what went on in the 1970s, the President's statement most certainly represented a major change in direction for the better.

Strategic Minerals: International Considerations

by William Schneider, Jr. *

As Congressman Santini correctly points out in his paper, this topic suffers from a general lack of understanding in the public sector, and to a considerable degree as well in the Congress and the Executive branch. Indeed, the Congressman accurately identifies a number of problems that currently exist in the strategic minerals field. However, a good deal has changed since mineral issues were first addressed seriously in the mid-1970s. It is important, therefore, to take note of what has been accomplished, as well as of what needs to be done.

It is generally agreed in the Executive branch, in Congress, and among the interested public that access to low-cost raw materials is essential to our peacetime prosperity. It is also generally understood that there is an uncomfortably high degree of overlap between areas outside the United States which are both principal resource producers and areas of substantial political instability. Southern Africa, of course, is an obvious case. But Southeast Asia and Latin America provide additional examples where there is substantial political turmoil as well as major sources of raw materials, access to which remains high on the agenda of both the major industrial powers and developing countries. Perhaps one could describe this situation as sheer coincidence, but it may be something more than that, especially insofar as political instabilities are concerned. For there appears to be a dimension of Soviet strategy toward these regions that is directly associated with the denial of resources to the West and Japan.

This is not to say that resource denial is the only interest Soviet leaders have in many mineralized areas of the world, but it does constitute an important motivation for them to engage in efforts to disrupt political life in these regions. It is fairly obvious, moreover, that the ability to deny access to resources from such areas does not necessarily require abject political control; merely imparting turmoil to a region through the threat of violence or guerrilla conflict,

---

* Dr. Schneider is the Under Secretary of State for Security Assistance, Science and Technology.

for example, is generally sufficient to deter adequate investment in mineral development in the Third World, where the time horizon on investment returns has to stretch over many years, and in some cases, decades. Indeed, such threats of instability provide an unsuitable environment for private investment from any country. It seems fair to conclude, therefore, that the issue of destabilization deserves greater attention in discussions focused on the international dimension of strategic mineral policy. And while this factor was to a considerable degree ignored during the early- and mid-1970s, I know, having served in the Congress and more recently in the Executive branch, that the potential for political instability in the Third World is now at least considered, even by the most skeptical observers, as something that the Soviet leadership can count on as a bonus to its activities in resource-rich regions.

As a consequence, strategic minerals are now on the list of important considerations in our diplomacy and foreign policy, and that development, in itself, is an important change, compared to the situation a few years ago. Now, in looking at what has happened to U.S. policy, and at some of the more important bench marks in recent years, a clearer characterization can be attempted concerning the way in which minerals policy has been injected into the calculation of our overall national interest.

Among recent initiatives, one of the most important developments in the long term—although it is unlikely to have any near-term effects—has been the rejection by the United States of the Law of the Sea Treaty, owing to the inadequacy of that Treaty with respect to its sea-bed mining provisions. Of course, because of the economic circumstances associated with terrestrial mining, it is unlikely that sea-bed mining will be a major factor any time in the near future. However, the regime that was proposed under the Law of the Sea Treaty would almost certainly have denied private industry in the advanced countries access to the sea-bed minerals under any reasonable—that is economic—circumstances. Therefore, the Reagan Administration's rejection of the Law of the Sea Treaty, and specifically its sea-bed mining provisions, together with the support of Treaty rejection by Congress, has made a very substantial contribution to the evolution of minerals policy as a long-range instrument of U.S. national policy. This has been coupled with the development of an alternative regime for sea-bed mining that will, in the long term, provide a legal basis for sea-bed mining by major industrial countries which have the technology, capital, and capability actually to conduct such activity.

The results of Treaty rejection and the creation of alternative plans for underwater mining can be seen in a number of areas. First, it will provide a hedge for industrial countries against incentives that might arise over the next several decades to deny access to terrestrial mining, especially in Third World areas. Second, it will also discourage the formation of cartels and tend to encourage the maintenance of conditions for access to foreign source raw materials on the basis of market conditions, rather than on the basis of political conditions. The Law of the Sea, while it is not going to make an immediate difference in terms of foreign source dependence, certainly will make a long-term difference by assuring that the legal regime under which mining takes place will not be prejudicial to the interests of the developed countries.

An additional area of interest that reflects a policy intention to render strategic minerals an important part of America's national focus is the Reagan Administration's interest in the U.S. strategic stockpile. Such concern, of course, is not new. We have had legislation at least indirectly relating to a strategic stockpile going back to World War I. Modern legislation emerged around the time of World War II, and we have had a large stockpile since the 1950s. The problem is that during the course of the late 1960s and early 1970s there was a change in the way the stockpile was managed. In particular, there was a growth of interest in using the stockpile as an economic instrument to offset short supply situations in specific minerals rather than as a vehicle for assuring our strategic independence from foreign resource supplies in the event of a national security emergency.

This nearly quarter-century of neglect with respect to stockpile objectives changed under pressure largely from the Congress—including Congressman Santini, who had an important role to play in articulating the concerns of the Congress, as well as Congressman Bennett of the House Armed Services Committee, who has expressed a rather intense interest generally in the subject of maintaining an adequate stockpile. There is no doubt that the subsequent efforts at stockpile reform could have been more effective: In some cases, the level of resource imports purchased for the stockpile, for example, could have been higher. Yet the state of mind regarding the stockpile, or the change thereof, is really the important difference. All the resources that recently have gone into the stockpile, the change in stock management plans whereby excess material may be sold off, and the potential use of the proceeds from those sales to build up stocks in the most critical materials—all these initiatives have a great deal of merit. There is general agreement, of course, that the stockpile is not as large as necessary to support the kind of strategic requirements that are likely to emerge in a military contingency, especially under circumstances that require a protracted surge in supporting military contingencies. Nevertheless, the corner has been turned, and this is certainly an important development. So is the renewal of the Defense Production Act (DPA) and the legal authority contained in it to support the domestic mineral industry as a way of offsetting foreign dependence. Even though this particular provision as yet has had limited application, the fact that the government has this authority, and has, in effect, confronted the prospect of using it, is significant.

Certain of the macro-economic problems that have hindered us from doing as much as we might have liked need to be addressed. Nevertheless, it should be recognized first that the strategic stockpile issue is important, as are the ramifications that U.S. interest in strategic stockpiles has had internationally. A number of the major NATO countries and, more recently, Japan, have demonstrated an interest in establishing an admittedly modest, but nevertheless visible, stockpile. Although many of these countries preserve some degree of ambiguity about the purposes that their stockpiles would serve, the mere existence of a stockpile provides the "raw materials," so to speak, for international cooperation on strategic stockpile policy. One of my hopes is that in the course of the next several years the principal industrial nations will be increasingly willing to consider the prospect of a national stockpile of critical materials—whether or not the stockpile is truly national in the sense that it is owned by the government in question, or is perceived as such as a result of tax incentives or other economic incentives provided by the government to help promote the development of stocks held by private entities. The point is that some coordination in this area can go a very substantial distance toward diminishing the dependence on foreign source raw materials that Mr. Santini and others have noted. This could become one of the "up sides" of the severe problems that our indigenous mining industry is having, because one aspect of current macro-economic conditions is the fact that it is now probably an ideal time for any country, but particularly the United States, to build up its stockpile, given the high value of the dollar and depressed commodity prices. Commodities can be acquired at costs that have not been seen since the 1950s, and this may provide a suitable incentive for stockpile purchases during the second term of the Reagan Administration.

An additional point concerns the Administration's posture, which the Congress has generally supported, against cartels. The powerful prejudice against supporting international producer agreements has, in general, helped to maintain the economic viability of the international minerals industry. I would argue further that it provides an appropriate basis for long-term stability in the industry by rendering it more subject to market influences and less subject to political influences.

Yet a further point which is difficult to demonstrate in any overt way, but which reveals an important change, is that American policy toward mineral producing countries now explicitly requires that the issue of U.S. access to raw materials produced by the country or countries in question be included on the agenda of our diplomatic representatives abroad, both on a bilateral basis and in international forums. This is something that historically has not been particularly well received in the Department of State, and it is not simply a problem related to minerals. In general, economic interests have taken second place in the past to diplomatic interests. But for many countries, economic interests really are now the predominant linkages that we have with them, and continued access to raw materials is an important case in point. I know in my own case that when new U.S. ambassadors are going to mineral-producing countries, I generally have a conversation with them before they depart, and discuss their plans for raising the visibility of mineral supply questions in bilateral dealings. In general, the response to this approach has been favorable, which, in itself, is another important change over past practice.

My last point is in some ways an offshoot of observations already made, and it relates to policy in Southern Africa, without question the region where minerals policy encounters its most difficult test, given the complex political problems there. I do not want to try and argue in detail for current policy toward Southern Africa, but there are a few points that ought to be stressed. First, our general policy in Southern Africa of "constructive engagement" has been trying to accomplish two ends. One is to push the Soviet Union out of a position of dominant influence in Southern Africa, a position it began to acquire in the late 1970s. The other objective is to press for the forces of change in Southern Africa which would encourage the development of socio-economic policies that might help stabilize the politics of the region.

The difficulty is that during the course of the 1970s the influx of Cuban troops and Soviet advisers into the region raised the security anxieties of South Africa, and, as a consequence, contributed to a reluctance on the part of Pretoria to make changes in social policy which would lean toward long-term political stability. Thus, one of the policy interests that we now have is to try to drive the Soviets out through a variety of means, including support for the recent Nkomati Accord between Mozambique and South Africa, which seems to be headed toward a considerable measure of success, as the government in Mozambique, for whatever reason, has found it in its interest to remove Soviet and other East European advisers from key positions, and to try and bring back Portuguese and other non-Mozambican technicians to rebuild the economy. This is having a bonus effect elsewhere in the region, where a number of the mineral-producing countries are beginning to change their rhetoric and, more importantly, change their policies in a way that could lead to a stabilization of regional politics. Recent events in Mozambique also have had a beneficial effect in South Africa itself, so that Pretoria is now more confident of its security as far as being threatened by foreign forces in the region is concerned, and has taken the first tentative steps toward mitigating some of the harsher aspects of its social policy. As a result, the long-term outlook for stability in Southern Africa has improved somewhat, and this is all to the good for the West and Japan because of the highly developed state of the minerals industry in that region.

One of the negative dimensions of the international economy in the past few years has been the effect it has had on the domestic minerals industry. The strength of the dollar, for example, has cut the cost of foreign copper by nearly 50 percent, simply because of the change in the exchange value of the dollar. In addition, the relatively depressed state of the minerals industry in the United States and elsewhere has tended to reduce the incentives for investment in the industry, and hence slowed its modernization. At the same time, some of the major producers, particularly South Africa, have decided to capture more of the value-added component of the resources they produce by building up their own indigenous mineral processing industries, particularly in specialty metals. This is a trend that is likely to occur in other areas of mineral processing as well. Some of these developments are likely to change as the exchange value of the dollar comes down to a more reasonable and sustainable level, but the fact remains that macro-economic considerations have had a very damaging effect on U.S. mining, at least for the short term.

However, protectionism is not, except in a few special cases, likely to be the best way to solve the problem. While we certainly need a copper and copper-processing industry, worldwide developments in this industry may suggest that we do not need as large a processing capacity as we had when we produced virtually all of our copper for local requirements. This is the kind of issue various government organizations should study, especially the question of exactly how large an industry we really need, once some sort of global economic equilibrium returns. It is important to note as well that the depressed commodity prices have a two-edged effect. By discouraging domestic mining and processing, they tend to damage severely the domestic industry. On the other hand, reduced prices also have the effect of maintaining more raw materials in the ground as a non-processed, strategic reserve in some sense, and this factor also needs to be weighed in the equation, when we ask how much of our indigenous resources we want to draw down, as opposed to tapping less costly foreign sources.

Soviet Global Policy and Raw Materials

*by John R. Thomas**

Any assessment of the role of natural resources in the USSR's foreign policy has to address three key elements. One is the history of Soviet views on raw materials. Students of Soviet history know that, even before the Bolsheviks came to power, they had paid attention to the relationship between the Western capitalist home countries and their colonies. In fact, there is a—possibly apocryphal—Leninist assertion to the effect that the road to Paris runs through Peking and Calcutta. Certainly, this reflected very early Soviet thinking, prior to the seizure of power, that the way to undermine and overthrow capitalist governments was to tamper with or block their access to the colonies on which they depended for raw materials.

The second point to be noted is the enhanced Soviet assessment of the correctness of this position, because—at least as far as the Soviet leaders perceive—the changes that occurred since 1917, viewed in the global context, have opened up opportunities both to make the most of the Soviet Union's own vast resources and to tamper with the resources of others upon which the West is compelled to draw.

Finally, we have to address specific policies and actions by the Soviet leadership in light of the points made above.

Before doing so, it should be pointed out that, on our side, it is not clear that the West has a thought-through, long-range policy. Moreover, any policies that may have been fashioned are difficult to implement in a somewhat fragmented, pluralistic society. The USSR, on the other hand, displays great perseverance, even in the face of severe domestic and foreign problems, nor is it easily discouraged by temporary setbacks. It is an open question whether a Western society, which had experienced large-scale, destructive civil war, the shock of collectivization of its agriculture, the extensive purges, the devastating German invasion of World War II, and the persistent underlying economic problems throughout that period, would be able to pick itself off the floor and keep going with a foreign policy offensive. Moreover, while Soviet

*Dr. Thomas is a Senior Soviet Affairs Specialist with the Department of State.

policies may appear to be—and often are—counterproductive, it would be a mistake to assume this fact would necessarily deter or inhibit Soviet leaders, or even be deemed by them to be truly counterproductive, as viewed from a Bolshevik perspective. The latter includes the view that societal life is a history of class struggle and strife, and that on the way to ultimate Soviet victory, temporary reverses have been and will be encountered; Soviet leaders have to and will take any setbacks in stride because of their Bolshevik discipline and organization.

This perspective is vindicated, in the Soviet view, by developments to date. A handful of Bolsheviks seized power in a coup d'etat in 1917, and, some 68 years later, their successors are in control of what is—at least militarily—a superpower. In other words, the USSR now possesses global power and influence, and this factor has direct impact upon Soviet raw materials policy, which has to be assessed in a long-term, global context.

With regard to Soviet long-range planning, there is little doubt that strategic minerals play an important role, with one caveat: Soviet implementation frequently does not match planning. Were this ever to occur, it would constitute a serious obstacle to American policymaking. It should be recalled that the USSR and its institutions not only have five-year planning, but even twenty-year projections. These are not infallible, of course, and they are modified frequently; however, the very existence of a requirement for such projections compels the institutions concerned to think about overall Soviet goals and about their specific contributions toward meeting these targets. In other words, when we note that the Soviet leadership has engaged in long-term thinking and planning with regard to raw materials, it also means the involvement of its military and economic, state, and party institutions. This discipline and integrative approach contrasts, of course, with our own situation, as described in other papers.

There are many examples relating to the Soviet perception of raw material problems, both domestic and international, including Moscow's ability to influence developments. This signifies that the USSR is thinking both about enhancement of its own vast resources and about manipulation of these resources as critical margins on the world scene affecting the West. Obviously, if there were no turmoil or instability on the current international scene, Soviet-controlled raw materials would be far less likely to have potentially critical importance. However, in Moscow's view, enough has happened during the last decade—e.g., the 1973 oil crisis which affected the West so drastically—that it reinforces their own projections of future crises in the world market. In turn, this leads the Soviet leadership to believe that the USSR's raw materials can and will play an important supportive role in advancing foreign policy goals. For instance, with regard to energy, the Soviet overall extraction and production capability increased from 8.3 percent of the world's supply of energy, in the period 1926-1950, to 19.1 percent, in the period 1976-1980. Without going into the specifics (e.g., the production of coal, oil, natural gas), this Soviet progress means an increase of roughly 150 percent in the Soviet Union's global role regarding energy, a factor that cannot but influence Moscow's assumptions about the future and its own strategy vis-à-vis the West.[1]

The Soviet leadership, of course, has noted well the impact of the energy crisis of the 1970s on the West. Moscow, moreover, is forecasting that this crisis has by no means been resolved; in light of this factor, Soviet diplomatic, economic, and political policies and actions are being pursued on a global scale to exploit and capitalize on the West's problems, and thereby advance Soviet foreign policy goals.

Comparing the global reach of the USSR today, it is worth noting that early Soviet foreign policy, under Stalin, was relatively conservative and cleaved to the areas immediately adjacent to the Soviet Union, with the exception of financing some communist parties abroad. Khrushchev, however, radically changed this situation: Soviet foreign policy became more activist and truly global, by jumping Soviet influence from areas adjacent to the USSR into parts of the Middle East, Asia, Africa, and Latin America that earlier had never seen a Soviet presence, e.g., via military and technical aid and advisers, diplomats, and Soviet Navy port calls. This transformation was significant even if Khrushchev's implementation did not fully match his aims, i.e., his expansive statements outran Soviet capabilities at the time. In this regard, Brezhnev and his successors have done much to close the strategic credibility gap by a massive Soviet military buildup, while their execution of other aspects of Soviet policy has been less noisy and blatant. Nevertheless, Khrushchev deserves almost exclusive credit for having opened the way for expansion of Soviet influence in the Third World—a major source of the world's raw materials.

This was achieved essentially by making an important amendment to Stalin's concept, as applied to non-Soviet bloc nations, of "you are either with us or you are against us." This created problems in the USSR's relations with the "nonaligned." Khrushchev modified Stalin's approach by positing to the LDCs, "if you are against the imperialist West, that is good enough for us and makes us natural allies, at least to some extent." This change in Soviet thinking brought the Third World and its raw materials within the scope of Soviet global policy.

[1]A more detailed analysis of Soviet raw materials, current production, and future potential is contained in my study, *Natural Resources in Soviet Foreign Policy*, a forthcoming Agenda Paper to be published in 1985 by the National Strategy Information Center, New York.

Regarding implementation, it is only recently that we have become aware how deeply the Soviet bureaucracy has become involved, from setting up within their leading strategic institutes study and analysis units devoted solely to raw materials, to engaging in more specific research concerning the strategic minerals vulnerabilities of such developed countries as Japan—a nation the USSR regards as the most exposed of the industrially advanced noncommunist states.[2] In this connection, Soviet analysts now have developed indices of raw materials production and consumption which deal not only with the Soviet domestic market, but keep a running tally of developments concerning the rest of the world. It should be noted that such work is intended for use by the top leadership, and helps to shape and formulate its perspectives and actions, as occasionally becomes evident in its official public pronouncements. For example, several years ago, the then Politburo member Kirilenko, in a review of domestic and foreign policy, addressed the energy crisis and compared the USSR's favorable and the West's unfavorable situation; his speech reflected the thrust of the raw materials studies done by Soviet researchers.

Viewing the USSR's raw materials perceptions and policies historically, we should note that in the early years after 1917 the Bolsheviks were fighting for survival and had neither the time nor the energy to devote to implementation of the concept that the road to Paris runs through Peking and Calcutta. Even so, Lenin and Stalin made the first efforts in the 1920s to begin undermining the West's position in the Third World, including the convocation of various conferences of activists from colonial areas, such as the Conference of the Toilers of the East. However, other priorities and emergencies intruded to cut short the Soviet effort at the time. But in the post-World War II era, the key element in the resumption of Soviet strategy toward the Third World was the disintegration of Western empires, as a consequence of the war, which opened up opportunities for the USSR to capitalize on crises in the ex-colonial parts of the globe.

Today, Soviet strategy operates at various levels, from the politico-military dimension reflected in the USSR's support for "national movements and wars of liberation," to the more conventional diplomatic arenas in giving its support to the New International Economic Order proposed by the less developed countries. The Soviet leaders also have pushed their own version of such proposals, labeling it as the New Raw Materials Order: beginning with a radical change in high prices to be charged the West for LDC raw materials, it is directed to undermining relations between the Third World and the noncommunist advanced industrial societies, including Japan. As Moscow

[2] A detailed description and analysis of the Soviet perception of Japan's raw materials vulnerabilities, and of Soviet plans to exploit them, is contained in my forthcoming study, previously cited.

sees it, the West lost physical control of the LDC's raw materials with the post-World War II disappearance of colonial empires; now the USSR's objective is to break up what it labels as "neo-colonialism," i.e., the West's economic/trade ties with the LDCs and its untrammeled access to their strategic minerals.

Seen from this angle, and compared with the West's unchallenged domination of the Third World in the interwar era, the USSR has had some significant successes in promoting and capitalizing on the West's uneasy relations with the LDCs. Consequently, we should not be overly impressed with the currently fashionable view in the West on the supposed failures of Soviet policy in the Third World. To be sure, the USSR has suffered some setbacks—as in the case of relations with Egypt under the late President Sadat—but Moscow is nothing if not patient. Indeed, some of its representatives have said openly, "don't worry, the Egyptians will come back to us." Moreover, while the relatively close Soviet-Egyptian relationship of the 1960s may not be fully restored, Mubarak is attempting to re-establish ties with Moscow, illustrated by the resumption of full diplomatic relations and the return of Soviet technicians—perhaps as insurance, just in case the West weakens or fails to defend its interests effectively in the Third World. Thus, Soviet patience accepts temporary setbacks and Moscow is not deterred from continuing to build up leverage against the West. If, therefore, we cannot rely upon Soviet setbacks as being permanent, neither can we rest upon our own successes, which may be just as temporary.

In any case, we can count on the USSR's perseverance and continuity in using raw materials as part of its foreign policy, as illustrated by Soviet past and current actions. Regarding the past, we should note how the USSR exploited its own raw materials, either as an inducement or as a threat, to influence or affect developments in other countries. In the pre-World War II period, the most illustrative case was the Soviet use of their oil and other resources to ward off, or at least delay, Nazi Germany's attack in June 1941. In the post-World War II period, Soviet leaders have used raw materials and related technology as part of their foreign policy to exert pressure both within and outside the communist world. Examples include Soviet actions against Yugoslavia (after Tito's break with Stalinism in 1948); China (after the outbreak of the Sino-Soviet dispute in 1960); Israel (after the USSR re-oriented its foreign policy toward the Arab cause in the late 1940s); Ghana (after the local military overthrew Nkrumah in 1966); Finland (where the Soviet Union was trying to influence the outcome of the Finnish elections in 1958); Cuba (as the USSR became involved in a dispute with Castro over appropriate revolutionary tactics in Latin America in the late 1960s); and Egypt (because Sadat expelled Soviet military advisers in the early 1970s).

In the instances cited above, Soviet actions against erring nations ranged from interrupting deliveries of raw materials (most particularly oil) and resource-related technology, to abandoning mineral processing facilities under construction in midstream, or failing to initiate construction projects that were scheduled to begin, based on previous commitments and agreements.

The objectives of Soviet use of resources as political leverage have included attempts to bring down an established regime (as happened in the case of Yugoslavia in the 1940s), efforts to wreck a nation's economy (China in the 1960s), calls for policy changes (Cuba in the 1960s when the USSR reduced oil shipments), and measures to influence the outcome of political elections (as in the case of Finland in the 1950s).

Soviet continuity in resorting to the raw materials "weapon" is illustrated in the Yugoslav case: Soviet leaders used this leverage against Tito in varying political circumstances, first under Stalin, then under Khrushchev. (The latter attempted to use this weapon, even though internally he was engaged simultaneously in the mid-1950s in "liberalizing" the Soviet system by his "de-Stalinization" campaign.)[3]

The role of raw materials in the USSR's foreign policy is also reflected in current words and actions. In this regard, the Soviet outlook has the following aspects:

- Soviet leaders *perceive* future mineral resource shortages outside the Soviet Union affecting the West and Japan, either because of possible physical depletion of such resources or because of political instability or turmoil in the LDCs.
- They are planning to exploit this situation, in part by using their own vast mineral resource potential, and in part by capitalizing on or instigating such turmoil and instability in other mineral producing areas, where and when the opportunity presents itself.
- To enhance their strategy as this relates to their own natural resources, Soviet leaders need foreign technology to develop their vast raw materials potential.

---

[3]For brief overviews of the Soviet use of raw materials as foreign policy leverage, see Marshall Goldman in the *Wall Street Journal*, November 23, 1981, and the *New York Times*, December 18, 1980; the London *Economist*, February 21, 1981; and John O'Sullivan, *Christian Science Monitor*, September 8, 1982. For a detailed examination, see Marshall Goldman, *Soviet Foreign Aid* (New York: Praeger, 1967); Milton Kovner, *The Challenge of Co-existence—A Study of Soviet Economic Diplomacy* (Washington, D. C.: Public Affairs Press, 1961); and Robert Freedman, *Economic Warfare in the Communist Bloc* (New York: Praeger, 1970.)

- They consider the Japanese a prime candidate from whom to obtain the needed technology, both because Japan has such technology and because, given Japan's dependence on imported raw materials, the Soviet Union could, in the Kremlin's view, serve in the future as a "stable source" of raw materials for Japan. Soviet leaders believe such a prospect of a reliable source will ultimately provoke Japanese interest because of intrinsic economic gain; and, at the same time, this prospect should also intimidate Japan into cooperating with the USSR, if Tokyo were confronted in the future by a threatened denial of access to Soviet resources in the face of Soviet-projected turmoil in alternative supplier countries.
- In addition to planning to exploit Japan's mineral source vulnerability, Soviet leaders are counting on applying, in the future, diplomatic and military pressure in order to arouse Japanese "interest" and acquiescence in meeting the USSR's technology needs.
- The use of all means, including resource exploitation as part of their foreign policy, is traditional with Soviet leaders; it stems from their historical view of the dependence of capitalist countries on their colonies and the likely problems the home countries would face if denied "colonial" (or neo-colonial, as the Soviet leadership now puts it) raw materials.
- Soviet views are supplemented by a well-documented past record of using raw materials in their foreign policy to pressure/intimidate and/or entice "target" countries.
- Given the Soviet perceptions and record to date, we can expect the Soviet leaders who are planning and are prepared to exploit the West's and Japan's resource problems to do so if given the opportunity.

The USSR's implementation of its natural resources policy and action on the perceptions noted above have already begun as demonstrated by the following developments:

- In general, the Soviet Union's current expansive post-World War II foreign policy has included attempts to control or influence sources of mineral supplies in the Third World.
- The Soviet Union also has tried to pressure recipient countries, both in and outside the Soviet bloc, by manipulating its own raw materials and related technology—i.e., to make them available or to withhold them.
- As part of specific targeting, a Soviet campaign to sensitize Japan to its long-run natural resource vulnerability and opportunities has already begun: on the one hand, by making thinly veiled threats of possible denial to Japan of access to Soviet Siberian and Far East resources and

of potential interruption of supplies from non-Soviet sources in the face of growing Soviet military capabilities, worldwide and regionally in East Asia; on the other hand, by enticing Japan with long-run prospects of access to Soviet raw materials "next door" and of great economic gain from Soviet purchases of Japan's technology to develop those resources. In this context, Soviet leaders and media have conveyed numerous messages to the Japanese that they should consider (a) their future raw materials needs and problems in light of the Soviet Union as a reliable, stable source, and (b) the effect of the future "correlation of forces in favor of the socialist world" on Japan's precarious raw materials position.

- To plan and implement their raw materials strategy in the years ahead, Soviet leaders have established a special research institute in Moscow to study raw materials in the global setting and to project future developments. This unit has produced studies projecting possible raw materials shortages or disruptions abroad which have already been used by the Soviet leadership.
- Because Soviet views on raw materials form an integral part of the broader Soviet geopolitical thinking, planning, and action (where and when the opportunity arises), Soviet actions in recent years have encompassed political, diplomatic, and military pressures when these were considered appropriate to the situation.

In noting Soviet perceptions and actions regarding the role of raw materials in their foreign policy, we leave open these questions:

- Will the Soviet leadership prove correct in its perception of future raw materials shortages and/or turmoil in LDCs?
- Will Japan, given Soviet assessment of its high vulnerability, prove amenable to future Soviet enticement or pressure?
- Will Soviet actions vis-à-vis Japan prove to be counterproductive?

These are issues for separate studies. But there should be no doubt that, from the Soviet viewpoint, raw materials have played and will play a key role in advancing Soviet interests and in undermining those of the West. We can ignore this Soviet "unity of thought and action" only at our own peril, given the long history of hostile Soviet actions and the stated Soviet objective to speed up "the historically inevitable decline of the West."

International Security Studies Program
The Fletcher School of Law and Diplomacy

Institute for Foreign Policy Analysis

Strategic Minerals Forum
September 17, 1984
Cabot Intercultural Center

Participants

Mr. E.F. Andrews
Vice President Materials & Services
Allegheny International
and Member,
National Strategic Material and Minerals Program Advisory Committee

Mr. Roland R. Carreker
Consultant
Materials Resource Analysis
General Electric Corporation

Mr. Bobby Cooley
International Relations Program
Tufts University

Ms. Bonnie J. Dancy
Program Consultant
American Program Bureau, Inc.

Mr. Richard E. Donnelly
Director, Industrial Resources
Office of the Under Secretary for Research and Engineering
U.S. Department of Defense

General Robert A. Duffy
Charles Stark Draper
Laboratory, Inc.

Ambassador Theodore L. Eliot, Jr.
Dean, The Fletcher School of Law and Diplomacy

Mr. Gregory Foster
Mobilization Concepts
Development Center
National Defense University

Lt. Col. Roger Garrett
USAF Research Associate
The Fletcher School of Law and Diplomacy

Professor Sheldon Glashow
Higgins Professor of Physics
Harvard University

Ambassador Edmund A. Gullion
Dean Emeritus
The Fletcher School of Law and Diplomacy, and
Chairman, Board of Directors
Institute for Foreign Policy Analysis, Inc.

Mr. Robert Johnson
Emergency Mobilization Coordinator
U.S. Bureau of Mines

Mr. Takashi Kawakami
Visiting Research Associate
Institute for Foreign Policy Analysis

Mr. Wallace E. Kirkpatrick
President, DESE Research & Engineering, Inc.

Professor Ernest Klema
The Fletcher School of Law and Diplomacy

Mr. Paul Krueger
Assistant Associate Director for Resources Preparedness
Federal Emergency Management Agency (FEMA)

Mr. Richard B. Levine
Deputy Director
Office of International Economic Affairs
National Security Council

Mr. Philip E. MacLean
President
Gorham International

Dr. Paul Maxwell
Committee on Science and Technology
U.S. House of Representatives

Mr. R. Daniel McMichael
Secretary, Sarah Scaife Foundation, Inc., and
Vice Chairman, National Strategic Materials and Minerals Program Advisory Committee

Ms. Frances Kramer Mesher
Consultant
Mineral Economics

Mr. John Mugar
Consultant to Norwood Group International, and
Former Chairman of the Board, Jewel Companies, Inc.

Richard D. Nethercut
The Fletcher School of Law and Diplomacy

Ms. Dorothy Nicolosi
Vice President and Treasurer
National Strategy and Information Center

Mrs. Jean Noble-Neal
Administrative Assistant
International Security Studies Program
The Fletcher School of Law and Diplomacy

Professor Joseph Nye
Kennedy School of Government
Harvard University

Dr. Charles Perry
Senior Staff Member
Institute for Foreign Policy Analysis

Professor John Perry
The Fletcher School of Law and Diplomacy

Professor Robert L. Pfaltzgraff, Jr.
Shelby Cullom Davis Professor of International Security Studies
The Fletcher School of Law and Diplomacy, and
President, Institute for Foreign Policy Analysis

Professor Uri Ra'anan
Director, International Security Studies Program, and
Professor of International Politics
The Fletcher School of Law and Diplomacy

Mr. Warren Richey
Staff Writer
The *Christian Science Monitor*

The Honorable James D. Santini
Bible, Santini, Hoy & Miller, and
Member, National Strategic Materials and Minerals Program Advisory Committee

Dr. John J. Schanz, Jr.
Senior Specialist
Congressional Research Service
Library of Congress

Dr. William Schneider, Jr.
Under Secretary for Security Assistance, Science and Technology
U.S. Department of State

Mr. Russell Seitz
Director, Technology Assessment
R.J. Edwards, Inc.

Professor Richard H. Schultz, Jr.
The Fletcher School of Law and Diplomacy

Mr. Robert Silano
Executive Director
Council on Economics and National Security

Mr. Robert Terrell
U.S. Senate Subcommittee on Energy and Natural Resources

Dr. John R. Thomas
Senior Soviet Affairs Analyst
U.S. Department of State

Ambassador Leonard Unger
The Fletcher School of Law and Diplomacy

Professor Arpad von Lazar
The Fletcher School of Law and Diplomacy

Mr. Dale Weiler
Senior Consultant
Arthur D. Little, Inc.

Mr. Robert Dale Wilson
Director,
Office of Strategic Resources
U.S. Department of Commerce

PERGAMON-BRASSEY'S
International Defense Publishers

List of Publications
in cooperation with the
Institute for Foreign Policy Analysis

Orders for the following titles should be addressed to:
Pergamon-Brassey's, Maxwell House, Fairview Park, Elmsford, New York, 10523; or to Pergamon-Brassey's, Headington Hill Hall, Oxford, OX3 0BW, England.

Foreign Policy Reports

Ethics, Deterrence, and National Security. By James E. Dougherty, Midge Decter, Pierre Hassner, Laurence Martin, Michael Novak, and Vladimir Bukovsky. June 1985. xvi, 91pp. $9.95.

Books

Atlantic Community in Crisis: A Redefinition of the Atlantic Relationship. Edited by Walter F. Hahn and Robert L. Pfaltzgraff, Jr. 1979. 386pp. $43.00.

Revising U.S. Military Strategy: Tailoring Means to Ends. By Jeffrey Record. 1984. 113pp. $16.95 ($9.95, paper).

INSTITUTE FOR FOREIGN POLICY ANALYSIS, INC.
List of Publications

Orders for the following titles in IFPA's series of Special Reports, Foreign Policy Reports, National Security Papers, Conference Reports, and Books should be addressed to the Circulation Manager, Institute for Foreign Policy Analysis, Central Plaza Building, Tenth Floor, 675 Massachusetts Avenue, Cambridge, Massachusetts 02139-3396. (Telephone: 617-492-2116.) Please send a check or money order for the correct amount together with your order.

Foreign Policy Reports

Defense Technology and the Atlantic Alliance: Competition or Collaboration? By Frank T. J. Bray and Michael Moodie. April 1977. 42pp. $5.00.

Iran's Quest for Security: U.S. Arms Transfers and the Nuclear Option. By Alvin J. Cottrell and James E. Dougherty. May 1977. 59pp. $5.00.

Ethiopia, the Horn of Africa, and U.S. Policy. By John H. Spencer. September 1977. 69pp. $5.00. (Out of print).

Beyond the Arab-Israeli Settlement: New Directions for U.S. Policy in the Middle East. By R. K. Ramazani. September 1977. 69pp. $5.00.

Spain, the Monarchy and the Atlantic Community. By David C. Jordan. June 1979. 55pp. $5.00.

U.S. Strategy at the Crossroads: Two Views. By Robert J. Hanks and Jeffrey Record. July 1982. 69pp. $7.50.

The U.S. Military Presence in the Middle East: Problems and Prospects. By Robert J. Hanks. December 1982. vii, 77pp. $7.50.

Southern Africa and Western Security. By Robert J. Hanks. August 1983. 71pp. $7.50.

The West German Peace Movement and the National Question. By Kim R. Holmes. March 1984. 73pp. $7.50.

The History and Impact of Marxist-Leninist Organizational Theory. By John P. Roche. April 1984. 73pp. $7.50.

Special Reports

The Cruise Missile: Bargaining Chip or Defense Bargain? By Robert L. Pfaltzgraff, Jr., and Jacquelyn K. Davis. January 1977. x, 53pp. $3.00.

Eurocommunism and the Atlantic Alliance. By James E. Dougherty and Diane K. Pfaltzgraff. January 1977. xiv, 66pp. $3.00.

The Neutron Bomb: Political, Technical and Military Issues. By S.T. Cohen. November 1978. xii, 95pp. $6.50.

SALT II and U.S. Strategic Forces. By Jacquelyn K. Davis, Patrick J. Friel and Robert L. Pfaltzgraff, Jr. June 1979. xii, 51pp. $5.00.

The Emerging Strategic Environment: Implications for Ballistic Missile Defense. By Leon Gouré, William G. Hyland and Colin S. Gray. December 1979. xi, 75pp. $6.50.

The Soviet Union and Ballistic Missile Defense. By Jacquelyn K. Davis, Uri Ra'anan, Robert L. Pfaltzgraff, Jr., Michael J. Deane and John M. Collins. March 1980. xi, 71pp. $6.50. (Out of print).

Energy Issues and Alliance Relationships: The United States, Western Europe and Japan. By Robert L. Pfaltzgraff, Jr. April 1980. xii, 71pp. $6.50.

U.S. Strategic-Nuclear Policy and Ballistic Missile Defense: the 1980s and Beyond. By William Schneider, Jr., Donald G. Brennan, William A. Davis, Jr., and Hans Rühle. April 1980. xii, 61pp. $6.50.

The Unnoticed Challenge: Soviet Maritime Strategy and the Global Choke Points. By Robert J. Hanks. August 1980. xi, 66pp. $6.50.

Force Reductions in Europe: Starting Over. By Jeffrey Record. October 1980. xi, 92pp. $6.50.

SALT II and American Security. By Gordon J. Humphrey, William R. Van Cleave, Jeffrey Record, William H. Kincade, and Richard Perle. October 1980. xvi, 65pp.

The Future of U.S. Land-Based Strategic Forces. By Jake Garn, J. I. Coffey, Lord Chalfont, and Ellery B. Block. December 1980. xvi, 80pp.

The Cape Route: Imperiled Western Lifeline. By Robert J. Hanks. February 1981. xi, 80pp. $6.50. (Hardcover, $10.00).

The Rapid Deployment Force and U.S. Military Intervention in the Persian Gulf. By Jeffrey Record. February 1981. viii, 82pp. $7.50. (Hardcover, $12.00).

Power Projection and the Long-Range Combat Aircraft: Missions, Capabilities and Alternative Designs. By Jacquelyn K. Davis and Robert L. Pfaltzgraff, Jr. June 1981. ix, 37pp. $6.50.

The Pacific Far East: Endangered American Strategic Position. By Robert J. Hanks. October 1981. ix, 75pp. $7.50.

NATO's Theater Nuclear Force Modernization Program: The Real Issues. By Jeffrey Record. November 1981. vii, 102pp. $7.50.

The Chemistry of Defeat: Asymmetries in U.S. and Soviet Chemical Warfare Postures. By Amoretta M. Hoeber. December 1981. xiii, 91pp. $6.50.

The Horn of Africa: A Map of Political-Strategic Conflict. By James E. Dougherty. April 1982. xv, 74pp. $7.50.

The West, Japan and Cape Route Imports: The Oil and Non-Fuel Mineral Trades. By Charles Perry. June 1982. xiv, 88pp. $7.50.

The Greens of West Germany: Origins, Strategies, and Transatlantic Implications. By Robert L. Pfaltzgraff, Jr., Kim R. Holmes, Clay Clemens, and Werner Kaltefleiter. August 1983. xi, 105pp. $7.50.

The Atlantic Alliance and U.S. Global Strategy. By Jacquelyn K. Davis and Robert L. Pfaltzgraff, Jr. September 1983. viii, 44pp. $7.50.

World Energy Supply and International Security. By Herman Franssen, John P. Hardt, Jacquelyn K. Davis, Robert J. Hanks, Charles Perry, Robert L. Pfaltzgraff, Jr., and Jeffrey Record. October 1983. xiv, 93pp. $7.50.

Poisoning Arms Control: the Soviet Union and Chemical/Biological Weapons. By Mark C. Storella. June 1984. xi, 99pp. $7.50.

National Security Papers

CBW: The Poor Man's Atomic Bomb. By Neil C. Livingstone and Joseph D. Douglass, Jr., with a Foreword by Senator John Tower. February 1984. x, 33pp. $5.00.

Books

Soviet Military Strategy in Europe. By Joseph D. Douglass, Jr. Pergamon Press, 1980. 252pp. (Out of print).

The Warsaw Pact: Arms, Doctrine, and Strategy. By William J. Lewis. New York: McGraw-Hill Publishing Co., 1982. 471pp. $29.95.

The Bishops and Nuclear Weapons: The Catholic Pastoral Letter on War and Peace. By James E. Dougherty. Archon Books, 1984. 255pp. $22.50.

Conference Reports

NATO and its Future: A German-American Roundtable. Summary of a Dialogue. 1978. 22pp. $1.00.

Second German-American Roundtable on NATO: The Theater-Nuclear Balance. A Conference Report. 1978. 32pp. $1.00.

The Soviet Union and Ballistic Missile Defense. A Conference Report. 1978. 26pp. $1.00.

U.S. Strategic-Nuclear Policy and Ballistic Missile Defense: The 1980s and Beyond. A Conference Report. 1979. 30pp. $1.00.

SALT II and American Security. A Conference Report. 1979. 39pp.

The Future of U.S. Land-Based Strategic Forces. A Conference Report. 1979. 32pp.

The Future of Nuclear Power. A Conference Report. 1980. 48pp. $1.00.

Third German-American Roundtable on NATO: Mutual and Balanced Force Reductions in Europe. A Conference Report. 1980. 27pp. $1.00.

Fourth German-American Roundtable on NATO: NATO Modernization and European Security. A Conference Report. 1981. 15pp. $1.00.

Second Anglo-American Symposium on Deterrence and European Security. A Conference Report. 1981. 25pp. $1.00.

U.S. Strategic Doctrine for the 1980s. A Conference Report. 1982. 14pp.

French-American Symposium on Strategy, Deterrence and European Security. A Conference Report. 1982. 14pp. $1.00.

Fifth German-American Roundtable on NATO: The Changing Context of the European Security Debate. Summary of a Transatlantic Dialogue. A Conference Report. 1982. 22pp. $1.00.

Energy Security and the Future of Nuclear Power. A Conference Report. 1982. 39pp. $2.50.

Portugal, Spain and Transatlantic Relations. Summary of a Transatlantic Dialogue. A Conference Report. 1983. 18pp. $2.50.

Japanese-American Symposium on Reducing Strategic Minerals Vulnerabilities: Current Plans, Priorities and Possibilities for Cooperation. A Conference Report. 1983. 31pp. $2.50.

The Security of the Atlantic, Iberian and North African Regions. Summary of a Transatlantic Dialogue. A Conference Report. 1983. 25pp. $2.50.

The West European Antinuclear Protest Movement: Implications for Western Security. Summary of a Transatlantic Dialogue. A Conference Report. 1984. 21pp. $2.50.

The U.S.-Japanese Security Relationship in Transition. Summary of a Transpacific Dialogue. A Conference Report. 1984. 23pp. $2.50.

Sixth German-American Roundtable on NATO: NATO and European Security—Beyond INF. Summary of a Transatlantic Dialogue. A Conference Report. 1984. 31pp. $2.50.

Third Japanese-American-German Conference on the Future of Nuclear Energy. A Conference Report. 1984. 40pp. $2.50.

Seventh German—American Roundtable on NATO: Political Constraints, Emerging Technologies, and Alliance Strategy. Summary of a Transatlantic Dialogue. A Conference Report. 1985. 36pp. $2.50.

Joint IFPA-ISSP Conference Reports

The U.S. Defense Mobilization Infrastructure: Problems and Priorities. A Conference Report (The Tenth Annual Conference, sponsored by the International Security Studies Program, The Fletcher School of Law and Diplomacy, Tufts University). 1981. 25pp. $1.00.

International Security Dimensions of Space. A Conference Report (The Eleventh Annual Conference, sponsored by the International Security Studies Program, The Fletcher School of Law and Diplomacy, Tufts University). 1982. 24pp. $2.50.

National Security Policy: The Decision-Making Process. A Conference Report (The Twelfth Annual Conference, sponsored by the International Security Studies Program, The Fletcher School of Law and Diplomacy, Tufts University). 1983. 28pp. $2.50.

Security Commitments and Capabilities: Elements of an American Global Strategy. A Conference Report (The Thirteenth Annual Conference, sponsored by the International Security Studies Program, The Fletcher School of Law and Diplomacy, Tufts University). 1984. 21pp. $2.50.

Terrorism and Other "Low-Intensity" Operations: International Linkages. A Conference Report (The Fourteenth Annual Conference, sponsored by the International Security Studies Program, The Fletcher School of Law and Diplomacy, Tufts University). 1985. 21pp. $2.50.